Teleports and
the Intelligent City

Teleports and the Intelligent City

Edited by

ANDREW D. LIPMAN
Pepper, Hamilton & Scheetz

ALAN D. SUGARMAN
Merrill Lynch

ROBERT F. CUSHMAN
Pepper, Hamilton & Scheetz

DOW JONES-IRWIN
Homewood, Illinois   60430

ISBN 0-87094-706-0

Library of Congress Catalog Card No. 86–70787

Printed in the United States of America

1 2 3 4 5 6 7 8 9 0 K 3 2 1 0 9 8 7 6

*This book, our second combined effort,
is dedicated to our parents:*
    *Edward and Martha Cushman,
    Maurice and Sylvia Lipman,
    and Nathan and Esther Sugarman.*

Preface

Teleports are combination telecommunication and real estate developments where the property owner or manager incorporates telecommunications and computer services and facilities (e.g., satellite dish antennas) primarily for long haul voice and data transmission, the cost of which can more efficiently be shared by a large set of users. Teleports are also becoming flexible communication nodes offering interconnection among satellite, fiber optic, microwave, and conventional land line networks, as well as interconnection among proprietary networks. Some teleports have established their own terrestrial networks and market them independent of other teleport services. Teleports are more than what the eye can see.

As real estate developments, developers package access to these technologies and afford both their tenants and off-site users within a particular region flexible access to services that otherwise would be too costly or difficult to secure individually. Developers can thereby establish previously untapped profit centers and boost the intrinsic value, profitability, and versatility of their properties. As developers become active participants in the quest to improve office system productivity, they will achieve greater visibility and a reputation for creating high-technology, employment-generating developments that can attract business investments and relocation to the region served by the teleport.

The teleport concept exploits technological innovations that require large-scale development. The developer who acts quickly can convert a parcel of land into a region's primary point of access to state-of-the-art data transmission facilities. In addition to on-site users, off-site users in numerous adjacent business districts can tap into the teleport via microwave and fiber optic cable links.

The teleport promises to provide less expensive access for small and medium volume users who might not generate enough demand for satellite and other efficient facilities, and for large volume users who prefer a "turnkey" operation rather than invest the time and effort necessary to develop a private telecommunications system. Teleport facilities equipped for multiple remote access also promise to solve problems for users in cities where existing telecommunications services prevent downtown construction and operation of additional microwave and satellite services (e.g., frequency congestion), or where available services do not satisfy particular needs.

Data transmission is a rapidly growing economic activity. Teleports can serve any enterprise which requires flexible, reliable, and economical access to telecommunications and information processing networks. Developers can exploit scale economies by affording access to both on- and off-site users who will realize cost savings by sharing facilities and services.

The following appear to be likely tenants or users of teleports:

1. Banks, insurance companies, multilocation corporations, high-volume data users or processors, and enterprises that seek to exploit telecommunications and information processing technologies to increase productivity and otherwise reduce costs, inefficiency, and travel.
2. Common carriers and resellers of telecommunications services.
3. Telephone order sales companies, for example, high volume WATS line users for direct customer sales.
4. Companies needing back-office data processing and archival storage facilities.
5. Institutions of higher learning.
6. Smaller businesses with significant telecommunications and data processing needs that can be met by pooled use of on-site capacity or by remote access to teleport facilities.

Among the likely revenue generating activities that a teleport developer can provide are:

1. Office space equipped with special enhancements to provide access to telecommunications and information processing services.
2. Multiple access to satellite earth stations and microwave facilities.
3. Resale of bulk satellite, private line, and WATS line capacity.
4. On-site land rentals for tenants' individual microwave towers and earth stations.
5. Technical and management services that expedite telecommunications and information processing accessibility and flexibility.

Development of a real estate parcel for teleport services involves cost, risk, political and regulatory issues, and promotional needs. A developer may wish to form joint ventures or partnerships with large telecommunications or data processing users or providers, or other equity partners. In recognition of the economic benefits accruing to a region, local authorities may assist in financing a teleport development by authorizing the issuance of industrial development bonds.

In the New York City teleport project, which aims to serve as much as 50 percent of the satellite transmission needs of this highly communications-intensive area, the New York Port Authority initially joined forces with the Western Union Telegraph Co. and Merrill Lynch & Co. for the development of a then projected $250 million project. Located on Staten Island, the New York teleport will have on-site office buildings and a field of satellite dish antennas that will be linked via fiber optic cables to the World Trade Center and other New York and New Jersey centers of commerce.

In Columbus, Ohio, a regional real estate developer joined with Compu-Serve, Inc., The Ohio State University, and the Chemical Abstracts Service Division of the American Chemical Society in a far less ambitious investment in a facility that will provide multiple audio, video, facsimile, and data transmission services via satellite and microwave facilities. Even in a medium-sized city like Columbus, the teleport developer has identified significant satellite telecommunications needs that have not been served locally. Other teleports are planned in Houston, Dallas, San Antonio, San Francisco, Washington, Chicago, and a host of other locations.

Some developers have enlisted large users as equity participants in major teleports to share the initial investment. By signing up a sophisticated major user, the developer will bolster the project and facilities interconnection with communications common carriers, which provide most transmission services between teleports and other telecommunications installations. The developer can then go about the task of educating and attracting other small tenants and users.

Building a teleport injects additional complexity to the conventional real estate development process.

1. *Frequency Clearances.* The developer must identify teleport sites that do not interfere with existing telecommunications facilities and are situated to provide functional satellite and microwave services without being so remote as to deter on-site office building tenants.

2. *Environmental Impact/Aesthetics.* Teleport transmission facilities (microwave towers and satellite earth stations) must satisfy energy radiation level limits and zoning restrictions that could entail added expense by requiring the developer to shield the teleport or require development in remote locations. If the facility adjoins residential areas, the developer may face additional expenses in building the system in a more attractive manner or constructing transmission facilities partially underground.

3. *Right-of-Way Acquisition.* Many potential customers may only want to lease teleport services without establishing a presence at the site. To make the teleport accessible to off-site users, for example, in the central business district, the developer may have to engineer high capacity microwave or fiber optic cable links as substitutes or cheaper alternatives to access provided

by the local telephone company. To establish remote cable access, the developer may have to acquire right-of-way leases from regional port authorities, railroads, local governments, or public utilities.

4. *Teleport Regulatory Status.* Depending on the configuration and services of the teleport and the ways in which common carrier services are interconnected, the developer can affect the potential for federal, state, or local regulatory oversight. Teleport developers should integrate legal and regulatory considerations as they conduct their preliminary economic, marketing, and engineering analysis.

This book includes a primer on telecommunications technology of teleports. Then, drawing on the experience of the well-recognized experts in this growing field, the authors discuss the planning, designing, regulating, operating, and marketing of teleports. Many of these topics are addressed more than once, but each time from the perspective of different professions and parties in interest. The reader can draw his or her own conclusions in evaluating these different viewpoints.

Chapter 1

In Chapter 1, Gerhard J. Hanneman of the ELRA group provides a general overview of teleports. By offering his initial definitions of the term as well as an introductory sketch of its history, rationale, implementation, and operation, Dr. Hanneman establishes a framework for the discussion which will follow in subsequent chapters. He also identifies and describes existing and planned teleports.

Chapter 2

For real estate developers and other professionals, the technology and terminology of telecommunications is potentially intimidating and is also a barrier to comprehending the potential of telecommunications-enhanced real estate. In Chapter 2, Glenn Pafumi of Dean Witter Reynolds and Dr. Stanley Welland of Merrill Lynch & Co. provide a primer on the basics of telecommunications technology. This chapter provides helpful background for many of the technical concepts that are developed in greater detail in subsequent chapters.

Chapter 3

Inherent to an understanding of the current status of the domestic satellite industry is some familiarity with the industry's history. In Chapter 3, Carl Cangelosi of RCA American Communications, Inc. provides a general overview of that history. Starting with the premise that the satellite business has shifted from having regulatory flexibility to being regulated by the marketplace, Cangelosi uses the subjects of satellites, earth stations, and rates to provide a helpful summary of the history of satellite regulation.

Chapter 4

Although teleports are the subject of this book, *teleport* is often a difficult term to define. As Kenneth Phillips of the Gulf Teleport indicates in Chapter 4, it is a term that describes a variety of high tech communications facilities. Expanding on Cangelosi's premise in Chapter 3 that the marketplace essentially regulates satellites, Phillips here suggests that the marketplace will also determine a teleport's specific form.

Chapter 5

After covering the essential concepts, history, and definitions of teleports in general, more specific types of teleports are discussed. Chapter 5 describes the largest teleport, the Staten Island Teleport, a joint venture of Merrill Lynch, Western Union, the City of New York, and the Port Authority of the States of New York and New Jersey. Robert Annunziata of Teleport Communications and Dr. Stanley Welland of Merrill Lynch provide an exciting description of such regional economic development projects as well as their prognosis for success.

Chapter 6

Not all teleports are as large as the Staten Island Teleport; medium-sized markets offer the possibility of a successful teleport venture as well. In Chapter 6, E. A. Eagan discusses one such venture—Central Florida Teleport—and then proceeds to offer practical advice about finding a workable teleport location and planning and marketing the teleport once a location is found.

Chapter 7

In Chapter 7, Miklos Korodi of the Ohio Teleport Corporation and Dr. Dennis K. Benson of Appropriate Solutions, Inc. discuss another medium-sized teleport, the Ohio Teleport. Korodi and Benson perceive this teleport as being distinguished from others by its attention to customer service, its continuing adaptability to evolving technology, and its solid adherence to the principle of end user corporate partnership. In describing these features, the authors add yet another component to the book's continually expanding definition of a teleport: its philosophical commitment to service.

Chapter 8

The last of a series of chapters which discuss specific teleports, Chapter 8 describes the Bay Area Teleport in California. Ronald H. Cowan here summarizes its planning and history, concluding that its combination of digital microwave and satellite facilities signals the advent of "the next generation of business communications."

Chapter 9

In Chapter 9, Thomas B. Cross of Cross Information Company provides an overview of teleconferencing, a major projected use of teleport capability. Teleconferencing enables people at multiple locations to communicate electronically. As Cross notes, teleconferencing is currently being integrated into the services provided by intelligent buildings and teleports. An understanding of this new technology is thus essential to the planning and designing of teleports and associated intelligent buildings, and Cross here gives an excellent introduction to the subject.

Chapter 10

In describing the Staten Island, Central Florida, Ohio, and San Francisco Teleports, the authors of Chapters 5 to 8 provide constructive suggestions regarding planning teleports in general. In Chapter 10, John Paul Rossie of Western Tele-Communications, Inc. offers a more general, systematic discussion of the planning process, including an analysis of common problems in planning and an overview of special design considerations. Rossie also raises some provocative questions regarding the consequences of world-wide networking and the challenge of finding common signaling protocols and standards of transmission.

Chapter 11

Chapter 11 provides technical specifics of some of the technology applicable to teleports, ranging from building cable and wiring systems to microwave and satellite technology. This chapter serves as an excellent companion to Chapter 10, as it addresses in more detail some of the issues of design and planning which Mr. Rossie introduced. Joseph Stern, Robert Weiblen, and T. J. Theodosios offer a thoughtful analysis of the various transmission facilities available to potential providers of high tech telecommunications in the intelligent city.

Chapter 12

Fundamental to the process of developing a real estate parcel for teleport services is the application process. As Mr. Cangelosi suggested in Chapter 3, regulatory considerations have been integral to the history of the satellite industry, and their importance continues today. In Chapter 12, coeditor Andrew D. Lipman and his colleague Emilia L. Govan describe the FCC regulatory process applicable to teleports and the various applications required in this process. The authors provide some initial guidelines for filing licensing applications for a few typical teleport facilities.

Chapter 13

In Chapters 3 and 4, Messrs. Cangelosi and Phillips noted the intrinsic relationship that exists between the marketplace on the one hand and the regulation and form of a teleport on the other. In Chapter 13, David Gourley of COMSAT International Communications, Inc. restates and enlarges this premise with a discussion of the relation between world economics and the development of international teleports. Beyond this discussion, Mr. Gourley here also provides a lucid examination of the distinctive characteristics, history, policy considerations, and potential problems of international teleports, echoing J. P. Rossie's questions in Chapter 10 about the consequences of worldwide networking and a "global village."

Chapter 14

Having provided an overview of the international satellite arena, the book next provides a series of chapters addressing several specific issues involving domestic satellites. In the first of these, Chapter 14, John N. Lemasters, president of Continental Telecom, and others, have written a succinct and clear description of satellite technology and how it is being applied, through facilities such as teleports, to provide satellite networks for real estate developers and corporate communications users.

Chapter 15

In Chapter 15, Dr. Joseph S. Kraemer of Touche Ross & Co. defines the nature of bypass and its effect on local telephone companies. A peculiarly domestic issue, bypass has particular relevance to teleports because local telephone companies consider them to be potential sources of the problem. In his discussion, however, Dr. Kraemer challenges this premise, concluding that the benefits that teleports may be able to provide to these companies outweigh the harm resulting from bypass.

Chapter 16

Teleports are only as viable as their long distribution network. Chapter 16, by Carl Gambello and Mel Van Vlack of Manhattan Cable, discusses the role of local cable and fiber optic networks in the high tech building and teleport environment. Many readers will be surprised to learn that cable television carries more than old movies and basketball games, and that today cable TV networks are being used as vital components in corporate voice and data networks.

Chapter 17

In Chapter 17, G. William Ruhl of Bell of Pennsylvania and Randall C. Frantz of Bell Atlantic discuss an alternative to private cable, satellite, and fiber optic networks for providing advanced corporate digital services—the telephone company network. They describe the evolution of Bell's analog voice network to the high tech digital/voice/data/video network; that evolution is well underway today.

Chapter 18

Teleports are in most cases significant real estate developments. In Chapter 18, architects Piero Patri and Michael Harlock, coeditor Alan Sugarman, and Peter Valentine explore the real estate aspects of teleports.

Chapter 19

Teleports are destined to become a key component of world cities. Professor Mitchell Moss examines the existing telecommunications infrastructure in New York City, and how the New York Teleport will fit into this environment.

In conclusion, the editors have structured this book as a vehicle to permit a wide array of professionals to recognize and take advantage of the significant opportunities available from teleports.

Andrew D. Lipman   **Alan D. Sugarman**   **Robert F. Cushman**
Washington, D.C.   *New York*   *Philadelphia*

Editors

Andrew D. Lipman

Andrew D. Lipman is a partner and head of the telecommunications department of the national law firm of Pepper, Hamilton & Scheetz, Washington, D.C., where he specializes in telecommunications and administrative law. Lipman is a graduate of Stanford Law School and University of Rochester, where he was a member of Phi Beta Kappa. He was in the legal honors program at the Department of Transportation and served as a trial attorney in the office of the Secretary of Transportation. He is a member of the Federal Communications Bar Association and the Interstate Commerce Commission Practitioners Association. He is an author of numerous articles on telecommunications, and lectures frequently on the subject.

Alan D. Sugarman

Alan D. Sugarman is vice president and associate general counsel for the real estate and real estate financial services group of Merrill Lynch & Co. Prior to assuming that position, Sugarman served as the general counsel of Merrill Lynch, Hubbard Inc., the group's institutional real estate division. While previously engaged in private practice in New York City, he specialized in litigation, corporate and real estate law. Sugarman was also general counsel, Roosevelt Island Development Corporation and senior staff counsel, INA Corporation. He is engaged presently in the practice of real estate, finance, investment and securities law. In addition, he has considerable experience in computer and telecommunications law and applications and other areas relating to technology. He is a graduate of the University of Chicago Law School, where he was a member of the *University of Chicago Law Review.* He also holds a bachelor's degree in electrical engineering from Tufts University and is a member of the Eta Kappa Nu and Tau Beta Pi engineering honor societies.

Robert F. Cushman

Robert F. Cushman, a coeditor of this publication, is a partner in the national law firm of Pepper, Hamilton & Scheetz and a recognized specialist and lecturer on all phases of real estate and construction law. He serves as legal counsel to numerous trade associations and construction, development, and bonding companies. Cushman is the editor and coauthor of *The Construction Industry Formbook,* published by Shepard's, Inc.; *The Dow Jones Businessman's Guide to Construction; The John Wiley Handbook on Managing Real Estate in the 1980s, Representing the Owner in Construction Litigation,* and numerous other titles. Cushman, who is a member of the bar of the Commonwealth of Pennsylvania and who is admitted to practice before the Supreme Court of the United States and the United States Claims Court, has served as executive vice president and general counsel to the Construction Industry Foundation. He is a member of the International Association of Insurance Counsel.

Alan D. Sugarman, Andrew D. Lipman, and Robert F. Cushman are editors of *High Tech Real Estate* (1985), also published by Dow Jones-Irwin.

Contributing Authors

Robert Annunziata   (Introduction and Chapter 5)
Teleport Communications, Inc.

Dr. Dennis K. Benson   (Chapter 7)
Appropriate Solutions, Inc.

Carl J. Cangelosi, Esq.   (Chapter 3)
RCA American Communications, Inc.

Albert F. Caprioglio   (Chapter 14)
Communications Systems Development

Robert E. Catlin   (Introduction)
New York Teleport

Ronald H. Cowan   (Chapter 8)
Bay Area Teleport

Thomas B. Cross   (Chapter 9)
Cross Information Company
  Intelligent Buildings Corporation

E. A. Eagan   (Chapter 6)
Central Florida Teleport

Randall C. Frantz   (Chapter 17)
Bell Atlantic

Carl Gambello   (Chapter 16)
Manhattan Cable TV

David E. Gourley   (Chapter 13)
COMSAT International Communications, Inc.

Emilia L. Govan, Esq.   (Chapter 12)
Pepper, Hamilton & Scheetz

Dr. Gerhard J. Hanneman   (Chapter 1)
ELRA Group

Michael J. Harlock   (Chapter 18)
Whisler-Patri

Miklos B. Korodi   (Chapter 7)
The Ohio Teleport Corporation

Joseph S. Kraemer   (Chapter 15)
Touche Ross & Co.

John N. Lemasters   (Chapter 14)
Continental Telecom Inc.

Andrew D. Lipman, Esq.   (Chapter 12)
Pepper, Hamilton & Scheetz

Mitchell Moss   (Chapter 19)
*Graduate School of Public Administration,
  New York University*

Glenn R. Pafumi   (Chapter 2)
Dean Witter Reynolds

Piero Patri, FAIA   (Chapter 18)
Whisler-Patri

Kenneth A. Phillips   (Chapter 4)
Gulf Teleport, Inc.

William J. Rahe, Jr.   (Chapter 14)
American Satellite Company

John Paul Rossie   (Chapter 10)
Western Tele-Communications, Inc.

G. William Ruhl   (Chapter 17)
Bell of Pennsylvania

Alan D. Sugarman, Esq.   (Chapter 18)
Merrill Lynch

Joseph L. Stern   (Chapter 11)
Stern Telecommunications Corporation

T. J. Theodosios   (Chapter 11)
Integrated Systems Planning, Inc.

Peter Valentine   (Chapter 18)
Comsul, Ltd.

Mel Van Vlack   (Chapter 16)
Manhattan Cable TV

Robert E. Weiblen   (Chapter 11)
Integrated Systems Planning, Inc.

Dr. Stanley M. Welland   (Chapters 2 and 5)
Merrill Lynch & Co.

George R. Welti   (Chapter 14)
Consulting Engineer

Introduction

As the developers and managers of the leading operating teleport in the world, we are proud to have been asked to write a short introduction to this book, which was conceived and edited by Andrew Lipman, Alan Sugarman, and Robert Cushman. The Teleport concept was born from the union of economic necessity and technological innovation.

In an effort to stem the migration of jobs and industry from the New York Region, the Port Authority of New York and New Jersey, the City of New York, Merrill Lynch, and Western Union sought to develop an infrastructure that would rebuild the economy from a technological base, and provide New York businesses with the facilities they needed to support information age activities.

At the same time, there was concern that the region would not be able to capitalize on the emergence of new communications technologies. New York's airwaves were as congested as its highways and unless an appropriate location could be found for satellite communications, the region would be a mere stop on the way to satellite communications centers far away.

What emerged was The Teleport, which combined an access facility to satellites, a real estate economic development project, and a fiber optic distribution network serving the greater regional community.

It was an idea whose time had come. Throughout the nation and the world, public and private sector organizations embraced the concept, and it became clear that it had applications that extended far beyond the Hudson River.

There are now over 20 teleports in the United States, with another 20 under construction. More than 8 major international teleports are being constructed in other parts of the world.

Teleports are evolving to become the new gateways for all regional international and interstate communications services. Just as ports are the transfer point for goods and services for a region, so are teleports becoming the transfer points for information. And increasingly, information is the commodity upon which modern economies are being built.

But what is particularly exciting about participating in the development of teleports is that they are still evolving. There are probably as many definitions of teleports as there are teleports, and the definitions offered today may be different from those offered in the future. This book is about that evolution. It provides an admirable overview and a snapshot of what teleports are today with a glimpse of what they will be in the future.

Teleports will expand from being carriers of information to manipulators of information. In the future, they will provide protocol conversions and transmission speed conversions to allow information to be communicated more easily throughout the world. They will be the information hub for seaports, airports, World Trade Centers, and financial communities.

Although the exact shape of teleports of the future cannot be predicted, one thing is clear. Teleports will help implement the dream of a smaller world that the communications revolution has promised.

The editors and the chapter authors they have assembled are to be congratulated on this important and timely book.

Robert Annunziata
President and Chief Operating Officer
Teleport Communications

Robert Catlin
General Manager
New York Teleport

Contents

ing. Pulse Code Modulation. Bandwidth Requirements for Various Types of Signals. Analog versus Digital Communications: Why Digital? Transmission Systems: *Wire Pairs. Coaxial Cable. Terrestrial Microwave. Optical Fiber. Communications Satellite Systems.* Data Transmission Codes: *Baudot. ASCII. EBCDIC.* Modes of Transmission: *Half Duplex. Full Duplex. Asynchronous. Synchronous.* Modems.

ogy. Customer Relations. Uncertain Futures. Ancillary Services: *Teleconferencing. Other Services.* Uncertainty Principle: *Transmission Speed. Transmission Channels and Destinations. Growth. Partnerships.* The Future.

Estate Development. Bypass and the Business of Teleports. Appendix: Definition and Nature of Bypass.

Cable Companies and Institutional Networks (I-Nets). I-Nets and the Marketplace. The Role of Local Cable Networks. Data Transmission and Cable I-Nets. Voice Services and Cable I-Nets. Video Services and Cable I-Nets. Cable Types and Characteristics: *Twisted Pair. Coaxial Cable. Optical Fiber. Coaxial Cable versus Optical Fiber.* Urban Cable Television Services: *System Usage. Customer Uses. Cable Installation. Technical Advantages. Financial Options.* Benefits to the I-Net User. Effects of AT&T Divestiture. Integration of Services. Challenges to Institutional Networks. Cable's Unique Position.

Communications—Understanding the Network: *Components of the Public Switched Network.* The Evolving Network: *WATS Service. Private Line. Foreign Exchange Service. Centrex. Alarm Service. Data Communication.* Divestiture: *BOC Responsibilities. Equal Access to Long-Distance Services.* Teleports and Telehubs: *Philadelphia—The Telehub.*

Centralization or Decentralization. The Port of the Information Age. Teleport Planning. Who Will Teleports Serve? *Teleport Benefits. Potential Users of Teleport Real Estate.* Teleports as Extensions of Smart Buildings and Parks. Two Teleports. Public, Private, or Both. How Many Teleports Per Region?

Introduction. New York City's Information Sector: *New York as an International Information Capital. An Assessment of Telecommunications Systems in New York City.* New York City's Long-Distance Telecommunications Systems: *Long-Distance Fiber Optic Systems. Satellite Facilities in New York City. Teleport: A Public-Private Partnership. Satellite Common Carriers in New York City. Private Earth Stations in New York City. Public Communication Networks.* New York City's Intracity Telecommunications Systems: *Fiber Optic Communications within New York City. Cable Television Systems in New York City. Terrestrial Microwave Systems in New York City. Cellular Mobile Radio in New York City. Smart Buildings and Local Area Networks in New York City.* Telecommunications and Economic Development.

Teleports and
the Intelligent City

Chapter 1

Teleports: An Overview

Dr. Gerhard J. Hanneman

ELRA Group

Outline

INTRODUCTION AND DEFINITIONS

HISTORICAL AND TECHNOLOGICAL
FACTORS AFFECTING TELEPORT
DEVELOPMENT

THE EVOLUTION OF TELEPORTS

TYPES OF TELEPORTS

IMPLEMENTATION AND OPERATION OF
A TELEPORT
Option 1: Stand-Alone Teleport

Limited/interim teleport
Full teleport
Option 2: Real Estate-Based Teleport
Summary

TELEPORTS AS HUBS OF THE
INTELLIGENT CITY
Real Estate as a Driving Force
Teleports and Economic Development

EXISTING AND PLANNED TELEPORTS

CONCLUSION

INTRODUCTION AND DEFINITIONS

This chapter is an overview of the teleport concept. It explores the development of the concept and factors that have supported the growth of teleports. This chapter also contains a description of the two major types of teleports, "stand-alone" and "real estate-based," and the steps involved in their implementation. Finally, the chapter concludes with a discussion of the role of teleports in the telecommunications infrastructures of cities as it relates to economic development. Also included is a description of existing and planned teleports throughout the world.

The material in this chapter is based upon a series of planning, design, and engineering activities conducted by the ELRA Group for teleport developers in San Francisco, San Antonio, and eight other metropolitan areas. In addition, the ELRA Group has been at the forefront of telecommunications infrastructure planning as this relates to the intelligent city. Through planning activities in Salem, Oregon, Minneapolis–St. Paul, and other metropolitan areas, we have begun to appreciate firsthand the significance of teleports as organizing elements of larger, more sophisticated telecommunications infrastructures, which are so important to completing this nation's transformation from an industrial society to an information-based society.

The American Teleport Association (ATA) considers an entity a teleport if:

It provides multiple and equal access to all customers and satellites.

It provides local and regional interconnect capability.

It is capable of communicating with other teleports.

Some existing satellite uplink facilities may meet the ATA definition of a teleport, but the operators of these facilities do not utilize the term *teleport* to market or describe their services. In other cases operators of multiple satellite antenna facilities have embraced the teleport designation. Teleports, however, differ from dedicated satellite antenna facilities operated by common carriers, in that the latter only link to the carriers' own satellites, whereas teleports provide open access to all satellites.

In essence then, a teleport is a communications facility or center that switches voice, video, and data communications to and from destinations, primarily using steerable and frequency-agile satellite antennas. A teleport offers shared access to communications satellites and to other transmission media—such as fiber optic networks—for local, regional, national, and international communications in a protected radio frequency environment.

Most teleports provide, or plan to provide, end-to-end service (full connectivity). However, in some cases the local telephone company, cable TV operator, or other carrier makes the "last mile" connection to the customer.

It is important to note that teleports are new concepts in telecommunications and that their growth and proliferation depend upon a number of exogenous conditions. These conditions include state and federal regulation and telecommunications policy; terrestrial and common-carrier competition, particularly from the new fiber optic networks; technological competition, particularly from new small-aperture satellite antenna terminals; and the continued growth in demand for communications capacity. Significant and rapid technological advancements suggest that teleports will become more than mere satellite antenna facilities. In fact, teleports currently under development are being designed more as network control centers—large communications switching facilities—in which the type of carrier medium accessed is transparent to customers. These centers will interface with fiber, coaxial, and satellite carriers interchangeably. As such, these teleports and their network arrangements will ensure economical and competitive transmission capacity for future teleport customers.

HISTORICAL AND TECHNOLOGICAL FACTORS AFFECTING TELEPORT DEVELOPMENT

Teleports are a manifestation of changes in society, in technology, and in the telecommunications industry. The factors influencing teleport development include telecommunications deregulation and technological advances in the information and communications industries, dramatic decreases in the cost of the space segment for satellite transmission, increasing use of office automation and computer communications, and a heightened realization among government officials at all levels of the relationship between the telecommunications infrastructure and economic development.

All of these factors, combined with the pressing demand for network capacity by large and sophisticated telecommunications users, led to the formation of the first formally designated "teleport." A joint venture of the Port Authority of New York and New Jersey; Merrill Lynch, Pierce, Fenner & Smith; and Western Union Telegraph Company formed (and coined the term) *The Teleport* in 1982 to be located on 350 acres of future office park on Staten Island. The Teleport represents the new, state-of-the-art, ground-up type of teleport. This is to be contrasted with antenna facilities already in operation as satellite gateways, which may also be called teleports. The former, ground-up type teleports tend to be designed around digital transmission, whereas the latter tend to be designed around analog transmission, primarily video.

The telecommunications industry continues to undergo significant deregulation based on the government policy that the public will be better served through market competition. Key judicial decisions opened up the telephone equipment sales and long-distance service industries (interexchange/interLATA) to competition. The most dramatic development was the AT&T divestiture in January 1984. Divestiture put the local telephone companies on an equal basis with other suppliers in marketing communications services, other than regulated local telephone service. This effectively meant that a whole new class of communications services was designated as "nonregulated" and subject to marketplace forces. Since divestiture, companies in many metropolitan areas requiring high-speed (56K) circuits have been put on long waiting lists, and access charges are causing considerable increases in business telephone costs.

Advancements in information and communications technologies have also affected the development of the teleport industry. With advances in computer technology, communications had to keep pace due to networking requirements and the growth of office automation.

Such technologies as fiber optics became cost effective, and new radio frequencies were allocated for high-speed data applications (Digital Termination Service). Changes in satellite spacing requirements and the success of the space shuttle program facilitated the use of satellites.

The growth in demand for communications capacity in the nation's largest cities has created new problems. Networks using terrestrial microwave radio communications have increasingly interfered with certain types of satellite transmissions. Key antenna sites are in some cases either unavailable or prohibitively expensive. Concurrently, businessmen, urban planners, and economic development officials have begun to realize that further economic development may be inhibited if these limitations are not overcome and investment in the telecommunications infrastructure is not promoted.

THE EVOLUTION OF TELEPORTS

A teleport is analogous to an airport in many respects. Airports function as air travel centers where costly resources, such as runways and control towers, are shared among all of the air carriers within a given geographic area. Airports are surrounded by a protected air space, which is kept free of intrusions that could disrupt operations. Travelers have access to multiple carriers at one central location. Aside from efficiently serving the needs of both providers and users of services, land use is maximized for the purpose of air travel.

Teleports function in a similar manner. Satellite earth stations, nodes for fiber and microwave distribution systems, relay stations, and equipment are centrally located in an environment that is protected from electronic interference. Users may enjoy access to a number of common carriers and services through one central facility. Carriers likewise have access to a variety of local distribution systems to serve their customers. Like an airport, a teleport provides for efficient use of facilities. Various carriers may be located at the teleport site, providing for efficient land use. Shared use of microwave and satellite frequencies maximizes their value.

Teleports take advantage of user aggregation not only through the sharing of common communication facilities and services but also through the sharing of real estate facilities. These include land, buildings, towers, power, signal interconnect, and cable distribution systems. Teleport sites are selected based on natural or artificial barriers that block potential microwave radio frequency interference (RFI) to or from sources. The site is also chosen to have maximum visibility of domestic and/or international satellites. Sites are also chosen in or near an urban area or an area with many heavy users of communications facilities for cost-effective local connections.

The teleport concept reflects the desire of customers to find technologically advanced voice/data capabilities that can be easily accessed and utilized at reasonable costs. The result is a centralization of the point of contact with suppliers. This recreates the one-stop communications service once offered by AT&T adding the benefit of choices among a variety of services and advanced transmission technologies.

In addition to major end-users, teleports accommodate the needs of a variety of segments involved in the telecommunications industry. Carriers that provide service only to other carriers may locate equipment at the teleport site if the served carriers base their relay stations at the teleport. Carriers that provide end-to-end service may wish to locate their points of presence at a teleport site to gain proximity to customers in the area. The trend toward distance-sensitive pricing for local telephone service may make this a key alternative for carriers.

TYPES OF TELEPORTS

There are two basic classes of teleports: stand alone and real estate based. Stand-alone teleports are sometimes referred to as antenna farms. This type of teleport is a centralized site where communications equipment, linked to satellite and local terrestrial networks, provides communications capabilities to users within a given geographic area. None of the users is physically located on site; the users must obtain access to the teleport services through a regional distribution system. The teleport functions as a "gateway" to communications networks outside of the area. The teleport itself is the sole revenue stream in a stand-alone operation. Stand-alone teleports already exist in Washington, D.C., Chicago, New York, and several other cities. Existing stand-alone teleports are primarily video-only (analog) facilities with long-term contracts to provide services for the CATV or television industries.

Real estate-based teleports are located in or adjacent to business parks and offer additional telecommunications services to park tenants on a shared basis. Through a partitionable telephone automatic branch exchange (PABX), tenants share advanced telephone features and other communications services, in addition to the long-distance and regional services available through a stand-alone teleport. The teleport functions as an amenity to help attract businesses to the office park. Also, due to the proximity of these tenants to the teleport, the tenants are viewed as a natural traffic base for teleport services. Indeed, the primary advantage of the real estate-based teleport is the captive traffic base potential.

In a real estate-based operation, the teleport itself is only one element in a larger revenue stream that includes multi-tenant shared services (MTSS). The Bay Area Teleport at the Harbor Bay Isle Business Park in Alameda, California, is an example of a real estate-based teleport. The Bay Area Teleport derives traffic from its affiliate MTSS company, Harbor Bay Telecommunications.

A variation of the real estate-based teleport is a user- or investor-specific teleport. In this type of teleport, the voice/data/video traffic and telecommunications of one very large user provide an organizing basis for a teleport. That is, a specific user (or users) by itself provides sufficient traffic and hence the revenue floor to justify the development and operation of a teleport. Merrill Lynch traffic at The Teleport in New York is the best example.

IMPLEMENTATION AND OPERATION OF A TELEPORT

Several strategic business options should be considered in developing a teleport. The most appropriate option depends on the outcome of the first step, the feasibility analysis, which includes market research of demand and potential competition and the financial analysis of operating options. Options include a limited or full stand-alone teleport, or a real estate-based teleport that would combine MTSS and teleport services.

Option 1: Stand-Alone Teleport

As described previously, the stand-alone teleport provides communication services to businesses and carriers within a region. Such a teleport would not be affiliated with an office park. A stand-alone teleport can be developed to various levels of sophistication. A limited or interim teleport would initially lease space to carriers or private network users and offer the capability for expansion as demand for services increased. A full teleport would include the development of or integration with a regional terrestrial network, access to all satellites, and a full complement of communications services.

Limited/interim teleport. For a limited teleport, only the minimum equipment and common carrier facilities necessary to provide essential teleport services are installed. To create a limited teleport, an area protected from frequency interference and with adequate size is essential. Pads for either an earth station or a microwave antenna tower would require between 250 and 2,000 square feet. Each earth station or tower would also require 1,300 square feet for a building to house equipment.

A limited teleport must provide access to local distribution systems and backup electrical power. Sophisticated hardware and software systems are needed to support the communication services and to track billing, performance, and system maintenance. In addition, human resource demands may be considerable depending whether the teleport serves only carriers or end-users themselves. Key to the success of a limited teleport is the ability to establish and maintain a sizable traffic base.

A telecommunications carrier with a network and customer base already in place could make the transition much more easily than a new start-up operation. An operator could offer some form of inducement or compensation to entice a carrier to establish the teleport on the basis of a profit sharing formula.

Full teleport. A full teleport is one in which the operator provides customers with a variety of communications services, including:

Local telco or other bypass alternatives for last-mile service.

Interface with fiber optic, CATV, and microwave distribution systems for regional and last-mile service.

Satellite transmit and receive facilities.

Access to common carrier and value-added networks.

Switching services.

Videoconferencing facilities with national connectivity.

A full teleport, like a limited teleport, also requires an area with suitable size and with no terrestrial microwave interference. This option involves three phases of development. During the first phase, financing is secured, engineering conducted, FCC licensing obtained, and request for bids from contractors let. The second phase includes construction, finalization of network arrangements, and last-mile connections, as well as initial operation. The third phase includes switch installation, marketing, and the turn-on of enhanced services, such as a videoteleconferencing center.

Option 2: Real Estate-Based Teleport

The second option is a real estate-based teleport, a teleport located in or adjacent to a business park. The real estate-based teleport would offer teleport and other telecommunication services on a shared tenant basis.

A multi-tenant shared services (MTSS) development would create a natural customer base for low-cost, long-distance and special communications services. A local area network in the business park would deliver all video, voice, and data traffic to the teleport, which would serve as a gateway to outside networks. Levels of MTSS are either basic or enhanced—basic services are defined as voice services, and enhanced services are defined as voice, data, and video services.

Development of a real estate-based teleport would involve the three phases described previously for a full teleport, plus integration of MTSS services. The levels and types of MTSS services installed would depend on the total square footage of the development, tenant characteristics and demand levels, and market competition. This opportunity would not only provide users with economical and easy access to the long-distance carriers located at the teleport, but would also provide a revenue stream for both projects. Equipment, marketing and technical personnel, computer software and hardware, and physical plant could all be shared.

Integrated MTSS and teleport projects usually have different implementation schedules, based on user demand and real estate absorption. This allows the operator more flexibility in staging the development of the project. For example, even if the operator decided to install separate switching systems at each building, the master system could be designed to provide temporary shared tenant service to the building until the occupancy level or feature or traffic demand warranted a stand-alone system. This would defray the costs of the stand-alone system until the users could actually pay for the separate system. If any delays in build-out or leasing were encountered, the operator would have avoided these costly delays. The same would be true of support and technical personnel requirements.

Summary

Although the steps in developing a teleport are quite complex, the process begins with a feasibility analysis. As part of this analysis, careful consideration should be given to whether or not ownership and operation of the teleport will be by the same entity. In the ELRA Group's experience, the majority of teleports in the United States represent a partnership or joint venture between investors/real estate developers/entrepreneurs and experienced telecommunications facility/network operators who have competitive and marketing savvy.

TELEPORTS AS HUBS OF THE INTELLIGENT CITY

Teleports continue to generate considerable attention for officials planning a telecommunications infrastructure for the citizens of the intelligent city. Cities, counties, and states are starting to examine the critical nature of the telecommunications infrastructure and its relationship to roadway infrastructure planning, sewage and water infrastructure planning, and other transportation facilities' infrastructure planning. In this context teleports are becoming the hubs of the intelligent city.

Macro-level relationships between telecommunications and infrastructures variables have long been recognized. One of the more comprehensive treatments of these issues is found in *World Communications: Threat or Promise?* by Colin Cherry (New York: Wiley Interscience, 1971), which examines the overall relationship among such constructs as health, wealth, literacy, development, social stability, and the use of telecommunications. Prior to the Cherry work, such scholars as Wilbur Schramm examined the mass media's role in disseminating information and promoting national development. Noted sociologist Daniel Lerner is his seminal book, *The Passing of Traditional Society,* examined the relationship of media-generated literacy and development in the Middle East. On a more formal basis, the International Telecommunication Union in Geneva has for at least two decades produced a series of reports examining and analyzing the relationship between telecommunication structures and economic well-being. More recently, the Committee on Telecommunications of the National Academy of Engineering as well as the RAND Corporation examined the relationship between communications technology and urban improvement.

All of these and other efforts suggest that the modern presumption has been that more and better telecommunications is in some way good because more access to information will generally lead to something better. Given a society that values information and information seeking, and in which institutional and economic leadership is frequently based upon the timeliness and validity of the information at hand, there is an imperative to critically analyze and develop the telecommunications infrastructure.

A teleport is particularly significant in light of three phenomena affecting society and cities:

1. Continuing competitiveness and deregulation of the telecommunications marketplace are allowing services, products, and networks to be developed at ever-increasing paces. Computers are being designed to switch at ever-increasing rates; computers the size of current desktop models that can store trillions of words instead of thousands of words are now only three or four years away from reaching the marketplace. Such computing capability, as generations of computer-literate students enter the work force over the next decade and increasingly rely upon computers for their information processing needs, will demand the development of high-capacity communications facilities. The current public switched network will be incapable of handling the explosion of traffic from networked personal computers.
2. The fact that the United States and other industrial nations are becoming nations of highly educated citizens who have been trained to use (and are comfortable with) information for their personal and business lives. As these citizens become a majority rather than a minority, the demand for information production, transfer, and processing capabilities will stretch the available telecommunication infrastructure.

3. Teleports bypass embedded interexchange networks. Cities in which citizens cannot have access to or be in receipt of information at a competitive rate with residents in other cities will be candidates for teleports. In many cases, because of the capital investment required in developing the telecommunications infrastructure, cities that can attract the facilities in the earliest phases of the information revolution will maintain dominance in information manufacturing in the future.

Real Estate as a Driving Force

The real estate industry in the United States is one of the driving forces in acknowledging teleports and telecommunications as an essential infrastructure element. This is because within the past three years teleports—and particularly shared or multi-tenant telecommunications services—have become a significant amenity and profit center for real estate developers. As communications intensive high tech industries become sought after by cities for their economic impact, realtors have embraced the teleport concept literally as a means of responding to local political needs and as a leasing inducement.

Teleports and Economic Development

Telecommunications infrastructure planning is key to future economic development of the city. As businesses and citizens begin to organize around the production of information, as opposed to the production of industrial goods or agricultural products, the quality and extent of the telecommunications infrastructure will become as essential as the quality and structure of roadways and other infrastructure elements were in previous eras. Telecommunications infrastructure will ultimately become linked not only to information abundance, but also to the survival and economic well-being of citizens. Cities that invest in such telecommunication structures as teleports are ensuring their economic survival in ways comparable to communities in earlier times that invested in canals, and then later roadways and airports. Economic survival was related to the ability of the citizenry to "get goods to market." The 21st century equivalent of "getting goods to market" will be transmitting information to and from remote terminals and data bases. This suggests that the city of the future has a very real stake in the future of teleports.

EXISTING AND PLANNED TELEPORTS

There are 36 planned or existing teleports in the United States and eight internationally; the major ones are described on the following pages, and a tabular listing of all teleports is included at the end of the chapter. The majority of teleports are located in or near large metropolitan areas, where the demand for high-speed telecommunications is greatest. Multiple teleports exist or are planned for certain cities, including New York, Washington, D.C., Houston, and Chicago.

CA: Bay Area Teleport—Alameda/San Francisco

The Bay Area Teleport (BAT) is a real estate-based project located across the bay from San Francisco in Alameda. The office park will provide clients with shared telecommunications services, including easy access to teleport services. The initial teleport plans call for installation of four 13-meter, one 9-meter and one 3.5-meter transmit/receive (T/R) satellite antennas. The antenna farm is located some miles from the central control facilities and will be interconnected via means of an 11-GHz microwave link. The microwave hop is part of a larger 23-node system that will connect the major San Francisco Bay Area cities to the teleport.

BAT will provide customers with a wide range of communications services, including a videoconferencing facility. The teleport will also be host to two COMSAT satellite antennas for the provision of INTELSAT Business Services (IBS), to the Atlantic and Pacific regions.

CA: Pacific International Teleport (PIT)—Los Angeles

The Pacific International Teleport is owned and operated by Wold Communications (as operator) and Mitsubishi Corporation. While Wold Communications has operated satellite facilities since the early 1970s, the Pacific International Teleport was inaugurated in 1985. The facility is a stand-alone common carrier currently divided between two nearby sites (one will house the Intelsat-B Antenna(s) and the other is the ABC facility from which Wold leases space). The latter site is used for all domestic communications. Both locations are connected by microwave, but Wold is looking for a new site to relocate all facilities. International service at PIT is expected to begin operation in 1986.

Local and regional interconnection is provided by means of both temporary and fixed microwave systems owned and operated by Wold. There is also a hardwire cable interconnect from a nearby Pacific Telephone central office, providing access to AT&T's point of presence. Wold Communications also either owns or leases transponder capacity on a number of domestic satellites.

In addition to the Pacific International Teleport, Wold Communications maintains satellite earth station facilities in the Washington, D.C., metropolitan area as well as in the New York City area.

CO: Teleport/Denver, Ltd.—Denver

Sat Time, Inc., a national satellite brokerage outfit, is developing a teleport in the Denver area to be operational by March, 1986. Temporary facilities are currently available. The teleport will be a private carrier communications facility associated with a real estate development: the Denver Technological Center and the Meridian International Business Center. The teleport plans to initially install five satellite antennas for domestic communications. There is also an application pending for provision of IBS service.

FL: Central Florida Teleport (CFT)—Ocala

Operational since September 1984, the Central Florida Teleport (CFT) is a real estate-based project located on 3.7 acres of the 305-acre Ocala Airport Commerce Center. CFT is currently equipped with only a single C-band transmit/receive (T/R) satellite antenna. Space for at least six more antennas is available. An office building with full telecommunications service, including a shared telephone system with least-cost routing, word processing, a teleconferencing studio, and satellite access, is scheduled to open in the fall of 1986. Office and TV production space will also be available at the facility.

Local and regional interconnection is provided by an 11-GHz microwave system operated by AT&T. In addition, a 27-mile 2-GHz microwave link will be established between CFT and a hub located at the University of Florida at Gainesville. A portable microwave system will be used to provide point-to-multipoint service on the campus.

FL: South Star Communications, Inc.—Davie

South Star Communications, Inc., operates an international common carrier in the vicinity of Miami and Ft. Lauderdale, Florida. The teleport began operations in July, 1985, and currently hosts seven satellite antennas with plans for an 11-meter international antenna for the provision of IBS service.

The facility has an on-premises 200-foot tower with two 2.5 GHz and a 7 GHz capability. Interconnection will also be made to the local telephone company by means of two fiber optic loops being developed for the teleport by Southern Bell, also for interconnection to the AT&T point of presence. Other private microwave and additional mobile/transportable units are available to clients.

GA: Atlanta Teleport (SSS)—Douglasville

Satellite Syndicated Systems (SSS) operates satellite facilities on an eight-acre site in Douglasville, Georgia. The facility has been in operation since 1979, distributing video, audio, and subcarrier services. Five transmit/receive satellite antennas are operational, and plans call for installation of at least three more. A terrestrial microwave system provides local and regional interconnection.

GA: The Turner Teleport—Atlanta

Turner Broadcasting System, Inc. (TBS) formed a subsidiary, Turner Teleport, Inc., to operate a teleport in Atlanta, Georgia. Applications have been filed with the FCC for authority to provide international business services and video services between Atlanta and Western Europe. Services will include voice, video, and data circuits. Teleconferencing facilities will also be available. The teleport is planned to be operational by late 1986.

GA: UpSouth Corporation—Atlanta

UpSouth has been in operation since March, 1983. It was purchased by Satellite Gateway Communications, Inc., a subsidiary of Pacific Telecom, Inc., in July, 1985. The facility is a common carrier on two acres in the northeast sector of Atlanta. There are seven satellite antennas at the facility, and plans for three more, one of which will provide international IBS service.

The teleport has a 300-foot tower which provides for microwave transmissions for regional and local interconnection. In addition, hardwire and microwave links are leased from Bell South for access to its central office, providing an interconnect to AT&T's point of presence.

HI: Hawaii Loa Teleport—Honolulu

The Hawaii Loa Teleport is planning to locate on 135 acres of private land owned by Hawaii Loa College (8 miles from Honolulu). The planning for this facility is being handled by Tele-Pacific Communications. Startup date for this facility is unknown. It is likely that the Hawaii Loa Teleport will be a noncommon carrier type facility, possibly tied to a conference center being planned at the college.

Regional interconnection to the teleport will be provided by a combination of microwave, fiber optics (owned and operated by Hawaiian Telephone), and cable television interconnect.

IL: Teleport Chicago—Chicago

Teleport Chicago is a stand-alone teleport located on a 2.3-acre site north of Chicago. The teleport site contains five transmit/receive satellite antennas. The owner, Midwestern Relay, operates an extensive microwave network in the Midwest.

IL: Chicago . . . ETC—Chicago

Chicago . . . ETC has been operating since 1981, providing mostly analog video transmission services. The site has three transmit/receive satellite antennas and six receive-only antennas. An additional 15 satellite antennas could be accommodated at this location. United Video, owner of Chicago . . . ETC, operates a terrestrial microwave link between downtown Chicago and the teleport in Monee, Illinois. Other microwave networks provide regional and local interconnections.

IA: The Iowa Teleport—Des Moines

The Iowa Teleport, owned and operated by the Heritage Teleport Corporation (a subsidiary of Heritage Communications), is scheduled to begin operations in the first quarter of 1986. This facility will be a noncommon carrier located on three acres just north of Des Moines. An additional video uplink located at Iowa State University (20 miles to the north) is also connected to the teleport site by means of microwave.

Regional interconnection to the teleport will be provided by Heritage's CARS band microwave system, as well as a combination of cable television interconnect (also owned by Heritage) and Northwestern Bell's and Iowa Power Company's (separate) fiber optic systems.

MA: Boston Teleport—Belmont

NET/PORT, Inc., has announced plans to develop a teleport in the Boston area. The facility is scheduled to come on-line in spring of 1986. There are plans to provide international IBS service as well as the full array of domestic communications services.

MO: Kansas City Teleport—Kansas City

The Kansas City Teleport, owned by Taft Television and Radio Company, has been in operation since 1981. It is located on a 15-acre site and operates three transmit/receive satellite antennas. The teleport currently provides for the transmission of analog video services; transmission of voice and data services will soon be available. Local and regional interconnection is provided by means of microwave and cable television institutional network facilities.

NY: The Teleport—Staten Island

The largest teleport project to date is located on Staten Island in the City of New York. Project partners are The Port Authority of New York and New Jersey, The City of New York, and Teleport Communications (a partnership of Merrill-Lynch Telecommunications and Western Union Communications Systems). Construction began in August 1983. The project includes an antenna farm and an office park with a shared telecommunications system. There are plans for 17 satellite earth stations for domestic communications. International communications will be transmitted by an on-site Comsat earth station providing the DIGITAL EXPRESS and Intelsat Business Service (IBS). Regional interconnection will be accomplished by a fiber optic network linking major cities in the area.

The fiber optic network will link the teleport to Manhattan, looping through the city and connecting such locations as the World Trade Center, the Empire State Building, and the Fisk Building. Cable extensions will tie the teleport to Astoria Studio in Queens, Metro Tech in Brooklyn, and the New Jersey cities of Princeton, Newark, Jersey City, and Somerset.

The Port Authority, which is responsible for the teleport land development, is negotiating with companies interested in locating large-scale offices— 150,000–200,000 square feet—at the 350-acre teleport site.

Teleport Communications has installed two 11-meter satellite antennas and leased land to Comsat for its IBS antenna. Negotiations are under way with other prospective customers for leasing of land for private satellite antenna placement.

NJ: Satellite City—Carteret

Satellite City is an earth station complex on 26 acres located in Carteret, New Jersey. Two 10-meter transmit/receive (T/R) antennas, one 12-meter T/R antenna, and one 7-meter T/R antenna are currently in operation. There is microwave interconnection to and within New York City. Applications for two international earth stations have been filed at the FCC for direct access to the INTELSAT system. Clients of Satellite City include ABC, BBC, Bonneville, Cablevision, CBN, CBS, Group W, Hughes, SIN, WNEW–TV, and other large television interests.

NC: Capitol Earthbase—Raleigh

Capitol Earthbase is a stand-alone teleport in Raleigh, North Carolina. The teleport operates two stationary and one transportable uplink. There are plans for adding one more stationary and two more transportable antennas. The control system is connected to local telephone company lines and an array of microwave equipment. Large video/audio feeds are the primary application.

OH: Ohio Teleport Corporation—Columbus

The Ohio Teleport Association was formed in 1982 to develop a telecommunications center in Columbus with uplink and downlink capabilities. The founding shareholders included Ohio State University, CompuServe Incorporated, The American Chemical Society, and Ruscilli Realty Company. The Ohio Teleport will be a real estate-based facility, possibly located at Ohio

State University. Two transmit/receive earth stations and a videoconferencing facility are planned to provide video, voice, and data services. The State of Ohio recently agreed to allow the Ohio Teleport Corporation to manage excess capacity on its state-wide microwave system.

OK: The Oklahoma Teleport—Oklahoma City

Ethereum Scientific Corporation and Rollins Communications have come together in a joint venture to operate the Oklahoma Teleport. The facility is a noncommon carrier on 25 acres located in the northwest sector of Oklahoma City. Service became operational in August, 1985. Local interconnection to the teleport will be made by means of private and leased microwave.

PA: Roaring Creek International Teleport—Mt. Carmel

The Roaring Creek facility is actually two separate communications hubs. The domestic and international video is handled by Roaring Creek International Teleport (RCIT). It is located on land leased from Comsat at the latter's Roaring Creek facility (Comsat's largest operation). RCIT provides domestic video turn-around for those international video transmissions handled by Comsat.

RCIT provides the gateway for communications from ANIK (the Canadian satellite system). It is also a U.S. signator for the Morales Satellite (Mexico). In addition, RCIT (through its parent companies) operates a microwave network from San Diego, California to Nuevo Laredo and the Caliente Race Track in Mexico.

PA: Pennsylvania Teleport—Avoca

Ethereum Scientific Corporation and N.E.P. Communications have joined to operate a stand-alone, common carrier facility located at the Wilkes-Barre/Scranton International Teleport. The teleport began operations in September, 1985. The teleport currently provides for only video services.

TX: Dallas/Fort Worth Teleport (DFWT)—Dallas

The Dallas Communications Complex (DCC), located at the 12,000 acre Las Colinas Development, is the operations site for the Dallas/Fort Worth Teleport, Ltd. (DFWT). The real estate-based teleport has four 5-meter dishes, transmitting more than 200 hours of video per week. A 9-acre site, interconnected to the DCC with fiber optic cable, has been selected as an ideal location for the antenna farm.

Companies located at the Las Colinas Development include IBM, NEC, Xerox, and Allstate. Access to the teleport will be accomplished either using the park's own coaxial cable system or a fiber optic cable. The fiber may either be installed and owned by Southwestern Bell Telephone (SWBT) or be a private dedicated link.

Regional interconnection will be made through SWBT's regional microwave network. DFWT operates its own 6-GHz and 23-GHz microwave network between Dallas and Fort Worth. It is also likely that some private microwave will be installed for clients with large telecommunications demand. SWBT has further agreed to connect the teleport directly to its regional toll center for access to the interexchange carriers.

TX: Gulf Teleport—Houston

Gulf Teleport was formed in September 1983 by a group of local Houston investors. The teleport is scheduled to begin operation in the southern part of Houston in 1986, with full operation of 12 satellite transmit/receive (T/R) dishes scheduled for 1986.

The teleport will provide direct microwave links (18 GHz) to major users located in the Houston area. Other means of local distribution—such as DTS, coaxial cable, telephone lines, and SCA frequencies—might be used. In addition, Gulf Teleport will build a major terrestrial microwave backbone connecting the cities of Lake Charles, Port Arthur, Orange, Beaumont, Houston, Port Lavaca, Victoria, and Corpus Christi. The network will provide the regional interconnection to the teleport.

The teleport has announced plans to locate its own earth terminals at other teleports in Columbus, New York, San Francisco, and Midland.

TX: Houston International Teleport (HIT)—Houston

The Houston International Teleport (HIT) is operated by Satellite Transmission and Reception Specialists (STARS). HIT has in place six satellite antennas of various sizes to provide video transmission for television broadcasters. In addition, there are four transportable antennas. Regional interconnection will be achieved by fiber optic cable, microwave, and local telephone plant. HIT is building a 100-foot tower located at the Johnson Space Center that will be used for microwave interconnection between the teleport and the space center.

TX: Houston Gateway Teleport—Houston

General Satellite Services, Inc. (GSSI) intends to develop a teleport facility southwest of Houston in Dewalt, Texas. Unlike the other teleports being developed in the Houston market, GSSI's facility will be positioned as a common carrier's carrier. As such, GSSI does not consider itself in competition with other telecommunications facilities. Instead, GSSI will provide an alternative for extra capacity demand.

Washington International Teleport—Washington, D.C.

Washington International Teleport is a stand-alone teleport owned by Carley Teleport Communications. The teleport has been operational for six years. The two- to three-acre facility houses four transmit/receive, one transmit, and four receive-only satellite antennas; each antenna is capable of handling both Ku-band and C-band communications. Clients include large broadcast and CATV programmers. Transmissions are currently limited to video/audio (analog) signal processing. There are no videoconferencing facilities. A terrestrial microwave network (MetroNet), operational at 11 GHz and 23 GHz, connects the teleport to locations in Northern Virginia and Maryland.

The National Teleport—Washington, D.C.

The National Teleport is owned by a partnership of Videostar Connections, Inc. (of Atlanta), and Pyramid Video Communications. The distinguishing feature of the National Teleport is its downtown location, atop the National Press Building. Other teleports are located in more suburban areas to take advantage of low radio frequency (RF) environments. The National Teleport, however, will take advantage of the Ku-band technology (14/12 GHz), which is rarely used by terrestrial microwave carriers and relatively free of interference. Space will be available for up to eight earth stations. Regional interconnection will be made via fiber optics, coaxial cable, and microwave.

International Teleports

As with the development of teleports in the United States, advancements in centralized telecommunication technologies are underway throughout the world. The prime goal of international teleports will be to facilitate a standardized global communication network meeting the demands of changing information technologies, as well as the needs of international trade and transportation.

The following are planned international-based teleports:

Jamaica Teleport—Jamaica
Tokyo Teleport—Japan
Osaka Teleport—Japan
Yokohama Teleport—Japan
Teleport Hamburg—West Germany
Rotterdam Teleport—The Netherlands
Telecenter Amsterdam—The Netherlands
London Drydocks—Great Britain

TABLE 1–1
Domestic Teleport Status Overview

Location	Teleport	Operational	Construction	Planned
California				
San Francisco Bay	Bay Area Teleport	X	(X)*	
Los Angeles	Pacific International Teleport	X		
Colorado				
Denver	Teleport/Denver, Ltd.			X
District of Columbia				
Washington	Washington International Teleport	X		
Washington	The National Teleport		X	
Washington	D.C. Teleport	X		
Florida				
Hollywood	Telelink Communications			X
Miami/ Ft. Lauderdale	South Star Communications	X	(X)	
Ocala	Central Florida Teleport	X	(X)	
Orlando	Orlando Telecommunications Port		X	
Georgia				
Atlanta	Turner Teleport	X	(X)	
Atlanta	UpSouth	X		
Douglasville	Atlanta International Teleport (SSS)	X		
Hawaii				
Honolulu	Hawaii Loa Teleport			X
Illinois				
Chicago	Chicago . . . ETC	X		
Chicago	Teleport Chicago	X		
Iowa				
Des Moines	The Iowa Teleport	X	(X)	
Massachusetts				
Boston (Belmont)	Boston Teleport			X
Boston	Boston Teleport (Taft)	X		
Missouri				
Kansas City	Kansas City Teleport	X		
New Jersey				
Carteret	Satellite City	X		
New York				
New York	The Teleport	X	(X)	
North Carolina				
Raleigh	Capitol Earthbase	X		
Ohio				
Central Ohio	Great Lakes Teleport			X
Columbus	COM III	X		
Columbus	Ohio Teleport Corporation			X

TABLE 1-1 *(concluded)*

Location	Teleport	Operational	Construction	Planned
Oklahoma				
Oklahoma	The Oklahoma Teleport	X	(X)	
Pennsylvania				
Roaring Creek	Roaring Creek International Teleport	X		
Scranton	Pennsylvania Teleport			X
Harrisburg	Harrisburg Teleport	X	(X)	
Texas				
Dallas	Dallas/Ft. Worth Teleport	X		
Houston	Gulf Teleport			X
Houston	Houston International Teleport	X		
Houston	Houston Gateway Teleport	Postponed		
San Antonio	Texas Teleport			X
Washington				
Seattle	Vashon Island (Seattle Teleport)	X		

* (X) = An operational facility with significant construction remaining.
Source: © ELRA Group, 1986.

■   CONCLUSION   ■

The teleport industry is robust and growing, as this chapter has indicated. Teleport services represent the "state of the art" in telecommunications and are becoming the hub of the intelligent city while evolving to accommodate the ever-increasing demand for voice, data, and video communications on a national and international basis. Although the AT&T divestiture has fragmented telecommunications services, teleports and MTSS offer a return to the one-stop shopping formerly available from AT&T. However, although teleports offer numerous advantages to businesses, it is important to note that they are still a new concept in communications. Their continued growth will depend upon competition from the existing communications capabilities in an area, the involvement of long distance or common carriers, and the competitiveness of the real estate marketplace.

Other factors will contribute to continued teleport development. These include societal values, such as workplace ethic, and trends in education, such as the increasing computer literacy of high-school and college students. These students will compose the work force during the second half of this decade and into the 1990s. The prevalence of these abundant computer-literate workers, the increasing societal acceptance of the telecommunications/transportation trade-off, and the diffusion of the personal computer in households will shape the intelligent city.

The continuing transformation of American society from matter production to information production and the increasing acceptance of automated decision support systems (e.g., artificial intelligence and "expert" systems) will continually expand work force utilization of telecommunications access networks (gateways). Thus information seeking, the value of data bases, and the telecommunications systems will become valued assets of businesses in the intelligent city.

As was discussed earlier, other trends will affect overall teleport growth. First, the convergence of the computer and communications marketplaces suggests improved economies of scale in hardware, and possibly, improved economies in the processing side of network operations. As educated citizens become a majority in the industrial world, further demands will be placed upon information production, transfer, and processing and upon the available telecommunications infrastructure. Finally, as communications needs escalate, the need for control of telecommunications costs will drive businesses to erect private networks or seek network alternatives that are more cost effective.

One of the significant future opportunities for teleport operators will be the emergence of the internationally accepted Integrated Services Digital Network (ISDN) concept. ISDN is a concept whereby voice, data, and video will be digitized in 144-Kb data streams to and from the home.

Technologically, the reconstruction of our existing analog networks in order that they may support ISDN in the forthcoming decades will require massive capital expenditures. Teleports are uniquely positioned to be functional as large-scale local and regional "front-ends" for national and international ISDN traffic.

Author

Gerhard J. Hanneman, Ph.D.
President and Chairman, ELRA Group
San Francisco, California

Gerhard J. Hanneman is the president and chairman of the ELRA Group, a San Francisco-based telecommunications consulting and research firm specializing in teleport planning. ELRA was responsible for the engineering and business planning for the Bay Area Teleport in the San Francisco area.

Dr. Hanneman was formerly the director of research for Xerox's XTEN subsidiary, the forerunner of the Digital Electronic Message Service. Prior to joining ELRA, Dr. Hanneman was a consultant to major Fortune 25 companies, the FCC, and the U.S. House of Representatives on various aspects of communications technology and policy. He was a professor and director of the Center for Communications Policy Research at the University of Southern California. He is the author or co-author of four books and 30 articles on various aspects of communications, including *The Telecommunications-Transportation Tradeoff.* His M.A. and Ph.D. are in communications from Michigan State University, where he was elected to Phi Beta Kappa.

Chapter 2

A Primer on Telecommunications and Related Technologies

Glenn R. Pafumi

Dean Witter Reynolds

Dr. Stanley M. Welland

Merrill Lynch & Co.

Outline

BASIC COMMUNICATIONS CONCEPTS
  Converting Speech to Electrical
    Impulses
  The Telephone Transmitter and
    Receiver
  Modulation
  Multiplexing
    Frequency division multiplexing
    Time division multiplexing
  Pulse Code Modulation
  Bandwidth Requirements for Various
    Types of Signals
  Analog versus Digital Communications:
    Why Digital?

TRANSMISSION SYSTEMS
  Wire Pairs
  Coaxial Cable
  Terrestrial Microwave

Optical Fiber
  Optical transmitters
  Types of optical fibers
  Characteristics of optical fiber
Communications Satellite Systems
  Satellite frequency bands
  Satellite orbits

DATA TRANSMISSION CODES
  Baudot
  ASCII
  EBCDIC

MODES OF TRANSMISSION
  Half Duplex
  Full Duplex
  Asynchronous
  Synchronous

MODEMS

Source: Reprinted from Alan D. Sugarman, Andrew D. Lipman, and Robert F. Cushman, eds., *High Tech Real Estate* (Homewood, Ill.: Dow Jones-Irwin, 1985), © Dow Jones-Irwin, 1985.

The purpose of this chapter is to explain how voice, data, and video communications are transmitted over conducting media and through the ether. We have taken an approach that will first examine the basic transmission mediums in use today, from a pair of copper wires to the more complicated coaxial cable, microwave radio, optical fiber, and satellite systems. In addition, the difference between analog and digital communications is discussed as well as an introduction to bandwidth, multiplexing, and digital systems.

The intent of this chapter is to focus on the engineering and telecommunication fundamentals in order to prepare the reader for the discussion, later in this book, of applications and technologies.

BASIC COMMUNICATIONS CONCEPTS

In 1876 Alexander Graham Bell patented the first telephonic device that converted voice signals—sound waves—into electrical signals—electrical waves. Since that time, devices have been developed that convert other types of information, such as data, images, and moving pictures into electrical signals. In a communications system, basic information, regardless of type, is transformed into electrical information by a sending terminal, whether it be a telephone for voice, a computer-like terminal for data, or a teletype device for telegrams or record-type communications, such as telex. The electrical information, or signal, is modified (in technical terms, *modulated*) by a sending device into a form suitable for transmission at a specific sending frequency. The modulated signal is then transmitted, or more properly propagated, by either a wire, radio, or video broadcast system. Finally, the propagated signal is demodulated or transformed at the receiving end back to the original information signal and the original basic information.

Transmission starts with the modulated signal being connected to a wire pair, a coaxial cable, a waveguide, an optical fiber, a microwave radio relay system, or an earth station (which transmits signals to a satellite). Most basic signals are concentrated at low frequencies and have to be modulated, since their lack of isolation from interference from nearby signals of the same frequency and their propagation (transmission) characteristics make them generally unsuitable for transmission in their existing state. Baseband (unmodulated) transmission employs wire (wire pair) channels, such as those used for local telephone loops or circuits. Baseband is used for the great bulk of short-distance communications and accounts for a large part of local telephone transmissions. However, with radio and other forms of longer-distance transmission, and in order to get greater capacity out of certain existing transmission facilities, it is usually necessary for the basic information signal to be transformed to higher frequencies that are suitable for electromagnetic propagation and can use the total capacity of the transmission facility.

The reader should note that the term *baseband* as applied to local area networks has a somewhat different meaning, and in general applies to a network that has a single carrier frequency.

Converting Speech to Electrical Impulses

Sound and electrical impulses are transmitted through a medium in wave form. A wave has three basic properties: amplitude, frequency, and phase.

In the basic sine wave, shown in Figure 2–1, the amplitude refers to the height of the wave. The amplitude represents the property of loudness in sound or the voltage level of an electrical impulse. Frequency, measured in cycles per second, or hertz, represents the number of times per second a wave completes a 360-degree shift in phase. Phase, measured in degrees, or radians, is the slope of a wave at a given instant in time. Frequency concerns the property of pitch. A sensitive human ear can detect frequencies of from about 30 hertz (or 30 cycles per second), which is a deep bass tone, to about 20,000 hertz, which is a high-pitched squeal. Most good quality high-fidelity stereos will reproduce sound over that same bandwidth. Bandwidth is the range of frequencies between the upper and lower limits of a transmission medium.

Ordinary speech usually covers a frequency range, or bandwidth, from 100 to 8,000 hertz. Good quality speech, however, can be transmitted over a bandwidth of from 300 to 3,400 hertz. Therefore, the typical telephone voice-grade channel, or subdivision of a circuit, has been designed to transmit only those frequencies that are between 300 and 3,400 hertz. That is why a person's voice over the telephone can have a slightly flat sound.

FIGURE 2–1
Basic Sine Wave

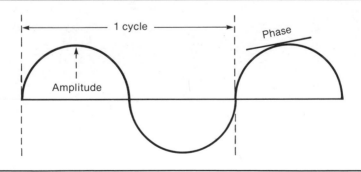

The Telephone Transmitter and Receiver

The telephone transmitter, or mouthpiece, is the ear of the telephone. In the transmitter, sound waves are changed into electrical waves or impulses to enable them to be sent over a telephone line to a distant receiver. The basic mouthpiece of the telephone consists of a diaphragm and a chamber filled with carbon granules.

Carbon has unique physical properties in that it is the only nonmetallic substance with electrical resistance that is low enough to allow its practical use as an electrical conductor and with a resistance that declines under pressure. This is shown in Figure 2–2.

When a person speaks into a telephone, the diaphragm in the telephone's mouthpiece vibrates at the frequency of the speaker's voice. The vibration alternately compresses and decompresses the chambers behind the diaphragm, varying the pressure of the carbon granules. The varying pressure on the granules varies the resistance of the granules. As a result of that varying resistance, the voltage level in the circuit that passes through the granules fluctuates. Therefore, the number of times the diaphragm vibrates determines the frequency of the electrical signal going through the transmitter. The varying resistance of the granules sets the voltage level and the amplitude of the electrical signal going through the transmitter. The telephone receiver is the speaker of the telephone. It reverses the process of the transmitter and produces sound waves from electrical impulses.

FIGURE 2–2
Telephone Mouthpiece

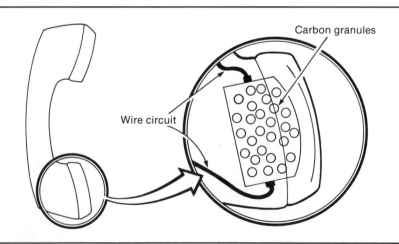

The frequency and amplitude of the electrical current running through the speaker causes the speaker to vary its vibration frequency (pitch) and amplitude (loudness) by the variation in the current. The vibration of the speaker reproduces the sound waves induced into the originating transmitter. The electrical signal (wave) is continuously changing in amplitude and frequency. It is referred to as an analog signal, in that it is an analog of the originating waveform, such as those produced by sound.

For most local communications, baseband transmission is used. That is, the transmission of electrical signals occurs at their originating frequencies. A pair of ordinary wires can easily handle a bandwidth of three kilohertz or 3,000 cycles per second. Thus the voice-carrying electrical signals (or current) that pass through the normal telephone are transmitted through the local telephone network to the receiving telephone unmodulated (unmodified).

The path of a typical phone call would be as follows. The electrical signals emanating from the telephone are carried over the local loop to the central office. The local loop consists of a pair of wires (usually referred to as a twisted wire pair) from the subscriber's phone to the local switching office. Subscribers not on party lines have their own dedicated pair of wires to this central office. In the central office, the call is connected to the party being called and then is sent to the called party via the called party's local loop, assuming the called party was connected to the same central office as the calling party. The transmission occurs in the baseband. If the call were being sent to a party connected to another central office, then baseband transmission would probably not be used over the entire transmission path. Instead, the call would be routed over an interoffice trunk (a circuit between two central offices) in a modulated form (see Figure 2–3). Obviously the local loop transmission would still be in the baseband. Only the trunk transmission

FIGURE 2–3
Modulated Transmission

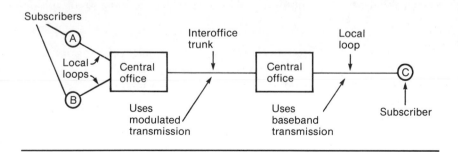

between central offices would occur in modulated form. The next section deals with modulation.

Modulation

One type of modulation is the process of superimposing a voice frequency (VF) channel on a carrier wave (a wave of constant frequency) so that a frequency shift can occur. For instance, a voice channel can be superimposed upon a carrier wave of 10,000 hertz (Hz) to shift its frequency range, or bandwidth, from a range of 300–3,400 hertz to a range of 10,300–13,400 hertz, as shown in Figure 2–4.

With demodulation, the reverse process is performed. The 10,300–13,400 Hz modulated voice channel band is superimposed on a 10,000 Hz carrier to recover the original 300–3,400 Hz voice frequency (VF) channel. The modulation process produces three components: the inserted carrier frequency and the two bands resulting from the superimposed frequencies. The sum of the carrier frequency and the voice channel frequencies produces the upper sideband. The difference of the carrier frequency and the voice channel frequencies produces the lower sideband.

The addition of a voice frequency (VF) channel to the 10,000 Hz carrier would produce the following signals:

Carrier + VF Channel   =

                                = *Carrier and sidebands*
10,000 Hz + (300 to 3,400) Hz = 10,300 to 13,400 Hz = Upper sideband
10,000 Hz   =   = Carrier frequency
10,000 Hz − (300 to 3,400) Hz = 6,600 to 9,700 Hz   = Lower sideband

Since the carrier frequency contains no information, and the lower sideband is redundant (has the same information as the upper sideband), the carrier wave and the lower sideband can be filtered out with the result that only the upper sideband is transmitted over the interoffice telephone line (or interoffice trunk). This is known as single sideband transmission (SSB). When the upper sideband is received, the demodulation process requires that the carrier frequency be reintroduced. This process produces the resulting sidebands:

Signal received   = 10,300 to 13,400 Hz = Upper sideband transmitted
Carrier frequency = 10,000 Hz − carrier frequency introduced
Resulting upper   = (10,300 to 13,400) Hz + 10,000 Hz
    sideband   = 20,300 to 23,400 Hz
Carrier frequency = 10,000 Hz
Resulting lower   = (10,300 to 13,400) Hz − 10,000 Hz
    sideband   = 300 to 3400 Hz

FIGURE 2–4
Modulation

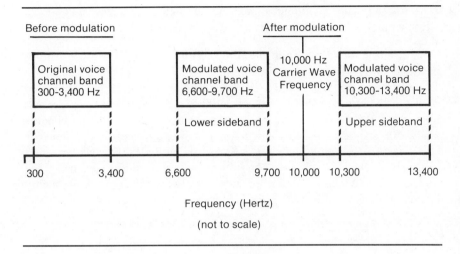

Before modulation | After modulation

| Original voice channel band 300-3,400 Hz | | Modulated voice channel band 6,600-9,700 Hz | 10,000 Hz Carrier Wave Frequency | Modulated voice channel band 10,300-13,400 Hz |

Lower sideband | Upper sideband

300   3,400   6,600   9,700 10,000 10,300   13,400

Frequency (Hertz)

(not to scale)

The final stage in demodulation is to filter out the upper sideband and the carrier wave. The signal recovered from the lower sideband, which has been received from the interoffice trunk, contains the original information signal from the originating 300 to 3,400 Hz voice frequency (VF) channel. It is then routed over local transmission facilities, in unmodulated or baseband form, to the called telephone.

Multiplexing

Multiplexing is the process by which more than one voice or data (or, for that matter, any communications) message can be transmitted or modulated over one physical communications transmission link or facility. There are three basic types of multiplexing techniques: space-division, frequency-division, and time-division multiplexing.

The simplest type is space-division multiplexing. Let us say we want to route 12 conversations, or telephone calls, over one transmission facility simultaneously. We could take 12 pairs of wires and group them into one large cable. That would be space-division multiplexing. Each conversation would occupy a separate wire pair or physical circuit (path). In actuality, wire pair cables are constructed to contain up to many hundreds of wire pairs. Such cables are placed in conduits under streets of cities or along roadsides and are used for subscriber local loop purposes and for short-distance interoffice trunks.

Frequency-division and time-division multiplexing are methods that have been developed to share one physical circuit, or path, among many conversations. Frequency-division multiplexing (FDM) uses the modulation technique previously described and is a natural in the world of analog signals and transmission (analog communications). Analog signals are signals, like sound, whose amplitudes and frequencies change continuously (see Figure 2–5).

Frequency division multiplexing. With frequency division multiplexing (FDM), many analog channels can be loaded onto one transmission circuit. As we said, a standard voice frequency channel generally runs from 300 to 3,400 Hz, or a bandwidth of 3,100 Hz (3.1 kilohertz). To prevent interference (crosstalk) among adjacent voice channels, the actual channel bandwidth, including spacing, used for a standard voice-grade channel is 4,000 Hz (4 kHz). Now, let's say we want to multiplex 12 voice channels over one transmission circuit. The 12 channels are multiplexed using the modulation technique just described. In the following example, for simplicity, assume single sideband (SSB) transmission is used.

Voice channel 1	= 4 kHz bandwidth
Voice channel 2	= 4 kHz bandwidth
.	
.	
.	
Voice channel 12	= 4 kHz bandwidth
Carrier frequency 1	= 100 kHz
Carrier frequency 2	= 104 kHz
.	
.	
.	
Carrier frequency 12	= 144 kHz
Voice transmission channel 1	= 100–104 kHz band
Voice transmission channel 2	= 104–108 kHz band
.	
.	
.	
Voice transmission channel 12	= 144–148 kHz band

Thus as one can see, if we have a transmission facility that has a bandwidth, or capacity, of from 100 kHz to 148 kHz, then we can send 12 voice channels over that one high-capacity (bandwidth) facility (see Figure 2–6).

Time division multiplexing. Time-division multiplexing (TDM) is an entirely different concept in the simultaneous transmission of signals.

FIGURE 2–5
Analog Signal

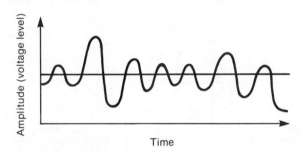

Whereas FDM is a natural for analog communications, time-division multi-plexing (TDM) is a natural for digital communications. Digital signals are signals, like data, whose amplitudes change in discrete levels. Digital information consists of bits (binary digits). In digital, the intelligence consists of a 0 or a 1, or an *on* or an *off* as shown in Figure 2–7 on p. 37.

In TDM the time available on a circuit is divided into slots, with each slot carrying one bit of intelligence, either a 0 or a 1. Only one channel occupies the circuit at any instant in time. The basic time-divided circuit is broken up into 24 channels. The first channel occupies the circuit long enough to enter a bit of information, either a 1 (a pulse or bit), or a 0 (which is the absence of a pulse or bit). Then the second channel enters a bit. This process continues up to the 24th channel and then starts all over again. In actuality, in telecommunications networks each channel enters eight bits of information before passing the circuit to the next channel.

In the basic DS–1 (digital signaling) rate, the collective 24 channels each contribute 8 bits of information or 192 bits. The entire process takes place in 1/8,000 of a second, or 125 microseconds. In Figure 2–8 on p. 38, channel 1 submits eight bits. Then channel 2 submits eight bits. The process continues until channel 24 submits its eight bits, and then the cycle begins again. All bits are sent on the DS–1 transmission line and are received in turn by channels 1 to 24 on the receiving end. The entire process is governed by a clock (at the top of Figure 2–8) so that the gate of channel 1 on the sending end of the transmission line coincides with the opening of the gate of channel 1 on the receiving end of the transmission line. The clock governs the opening and closing of all the gates for all 24 channels attached to the transmission line. (The use of a clock on both ends of the transmission line is known as synchronous transmission.) The process of sending 8 bits per channel for

FIGURE 2–6
Multiplexing

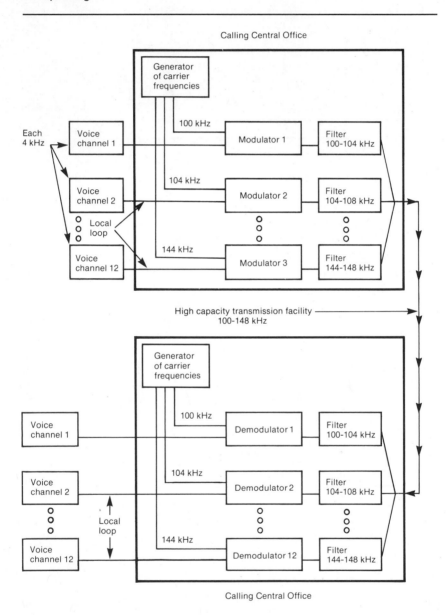

FIGURE 2–7
Digital Signal

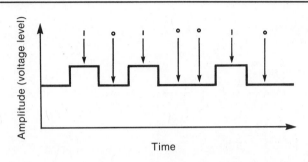

all 24 channels plus 1 bit for synchronization (totaling 193 bits) takes 1/8,000 of a second.

As shown in Figure 2–9, each frame consists of 193 bits,with 24 channels of 8 bits each (24 × 8 = 192 bits) plus a 193d bit used for synchronization. Note that in our example, only the first seven bits of each eight-bit word carry information, either a 0 or a 1, since the eighth bit of every word is used by the network for signaling and supervisory information. There are more efficient techniques employed today, but it would not serve the purpose of this chapter to describe them. As we said, the entire process or frame takes 1/8,000th of a second to complete and is repeated 8,000 times a second. This is how we arrive at the DS–1 signaling rate of 1,544,000 bits per second or 1.544 megabits per second.

8,000 frames/sec × 193 bits/frame = 1,544,000 bits/second

Time-division multiplexing (TDM) is ideally suited for digital signals. It permits many digit (bit) producing machines, like data terminals, computers, and so on, to share a single transmission line by interleaving the bit streams of more than one terminal over one transmission device. However, analog signals can also be sent over digital transmission lines. But to do so, they must first be converted to digital signals. The first step in the process of converting analog signals into digital form is *quantizing*. The method used to encode the quantized signals is *pulse code modulation.*

Pulse Code Modulation

With pulse code modulation (PCM), a telephone voice becomes a bit stream that looks like digital data. A codec, for *co*der-*dec*oder, converts the analog signal to digital form. Ironically, the trend in the communications industry

FIGURE 2–8
DS–1 Signaling Process

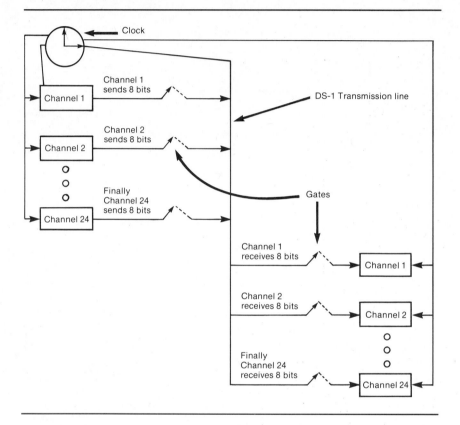

The process illustrated constitutes one frame which is repeated 8,000 times a second and produces the DS–1 signaling rate of 1,544,000 bits per second. A look at the signal going over a DS–1 transmission line is shown in Figure 2–9.

is via a codec to convert the more prevalent analog signals (such as voices) so that they may travel over digital lines. Whereas in the computer industry the trend is to convert less prevalent digital signals, such as data, via a modem (short for *mo*dulator *dem*odulator), to travel over analog telephone lines.

To convert an analog signal, such as a voice, to a digital signal, or pulse train, it is required that the circuit over which the analog signal travels be sampled (measured in amplitude) at periodic intervals. The simplest type of sampling (measurement) would produce pulses that are proportional to the amplitude of the sampled signal at the instant it is sampled. This process

FIGURE 2–9
PCM Frame

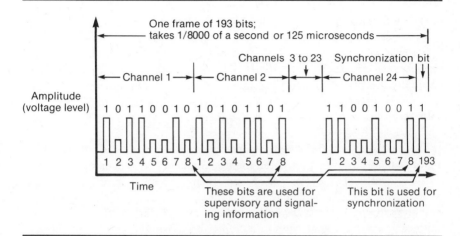

Note that the 0 bit is shown as a bit of lower amplitude than the 1 bit. However the 0 bit could have also been the absence of a bit or pulse while the bit the presence of a bit or pulse.

is called *p*ulse *a*mplitude *m*odulation (*PAM;* see Figure 2–10). The pulses in Figure 2–10 still carry their information in quasi-analog form, since the amplitude of each pulse can take on an infinite number of possible values from zero to the maximum amplitude of the wave passing through the circuit.

FIGURE 2–10
Pulse Amplitude Modulation (PAM)

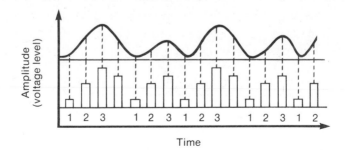

However since we want to digitize the voice samples, or the amplitude modulated pulses, a second process is employed that converts the PAM pulses into unique sets of equal amplitude pulses. The receiving equipment then needs to detect only whether a bit or a pulse is a 0 or a 1. The absence of a bit is interpreted as a 0 and the presence of a bit as a 1. The amplitude of each individual pulse is not relevant since, at least in theory, it is constant.

As we said, the amplitude of a PAM pulse can take on infinite number of possible values from zero to the maximum amplitude of the wave passing through the circuit. However with pulse-coded types of modulation, it is normal to limit the number of possible amplitudes to a limited set of discrete values. The process of dividing an infinite range of possible amplitudes to a finite set of discrete amplitudes is called quantizing. The process is illustrated in Figure 2–11.

The next question becomes how many samples are needed. It has been mathematically proven that if the highest frequency in a channel is limited to a bandwidth of X cycles per second, then a sampling rate of $2X$ cycles (samples) per second is sufficient to carry the signal and completely reconstruct it. If the sampling rate drops below twice the bandwidth of the signal being sampled, then the signal cannot be reproduced without distortion (noise). If the sampling rate is greater than twice the bandwidth, then redundant sampling occurs.

Since a voice-grade channel has a maximum frequency of 4,000 cycles per second (4 kHz), the sampling rate should be 8,000 samples per second. Indeed, when we discussed the frame in Figure 2–9, we said that the frame samples 24 channels in 1/8,000 of a second. Thus each voice-grade channel, sampled 8,000 times a second, produces 8,000 PAM pulses per second. Since each PAM pulse is converted into a 7-bit code, a 4 kHz standard voice-

FIGURE 2–11
Quantizing

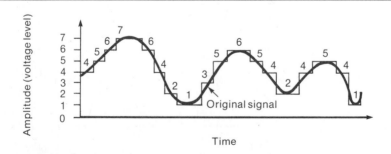

grade channel uses 8,000 times 7, or 56,000, bits per second (bps). An eighth bit is added to each sample for signaling and supervisory purposes, adding 8,000 bits per second (bps). Thus a standard voice frequency channel that carries voice, signaling, and supervisory information operates at 56,000 bps (for voice) plus 8,000 bps (for signaling and supervisory data), or 64,000 bps.

The DS–1 (*d*igital *s*ignaling) rate transmits 24 voice channels. Thus its frequency, or bit rate, is 24 × 64,000 bits per second (bps), or 1,536,000 bps. Since 1 bit is added to each frame for synchronization purposes, 8,000 bits per second are used for synchronization. Thus the DS–1 rate becomes 1,536,000 bps (for 24 voice channels) plus 8,000 bps (for synchronization), or 1,544,000 bps (which equals 1.544 megabits per seconds—1 megabit equals 1 million bits).

Bandwidth Requirements for Various Types of Signals

Thus far we have based discussions on the standard voice frequency channel. This has been done for the following reasons: (1) The predominent use of today's telecommunications links is for voice communications; (2) most of the design of today's telecommunications systems have been based on voice frequency channels; (3) voice will continue to account for the great bulk of the usage of telephone lines for many years to come; and (4) most future telecommunications systems will be designed considering the constraints of voice transmissions first. However, voice is not the only type of communication that can be sent down a telecommunications link. Telegraph signals, data, low-fidelity sound, high-fidelity sound, still images (facsimile), and moving pictures (Picturephone and television in both black and white and color) can also be transmitted over a telecommunications medium. All of the signals can be transmitted in either the analog or digital mode.

However, different types of signals require different amounts of bandwidth, or capacity, to transfer the information they carry. Table 2–1 indicates the amount of bandwidth that is required to carry each type of analog signal in the analog mode. Next to the analog bandwidth requirement is the sampling rate per second needed to convert the analog signal into a stream of pulse amplitude modulation (PAM) pulses. Regardless of the signal source, a sampling rate equal to twice the bandwidth of an analog signal will permit an analog signal to be reproduced accurately. Next to the sampling rate is the number of discrete quantizing levels required to make the quality of the signal acceptable and to reduce quantizing noise to tolerable levels. (A voice frequency circuit's PAM pulses are quantized into 128 or 256 discrete levels.) The following column indicates the encoding scheme used to convert the PAM pulses into a digital bit stream of 0 or 1 (on or off) pulses, or pulse code modulation (PCM) bits. (A voice PAM pulse is converted into a seven- or eight-bit code to permit and identify each of 128 or 256 discrete quantizing levels—$2^7 = 128$; $2^8 = 256$). An n-bit encoding scheme would permit 2^n discrete quantizing levels. The last column in the table indicates the bit rate required to carry the particular analog signal in digital form. At the bottom of Table 2–1 some digital signals and speeds are included for comparative purposes.

Analog versus Digital Communications: Why Digital?

Analog communication generally utilizes less bandwidth (capacity) than digital communication of the same information. A standard analog voice frequency circuit has a bandwidth of 4 kHz, whereas a digitized voice signal can require 64,000 bits per second of digital data, which can utilize up to 64 kHz of bandwidth. In addition, for the foreseeable future the bulk of communications will be provided for analog signals (voice and TV) and not data. So the question becomes: Why digital?

The Bell System introduced digital PCM communications, via the 24-channel T-1 carrier system, in 1962. Digital communications made economic sense in short-haul applications. AT&T had many wire pairs under city streets, and instead of using each of them to transmit one analog voice, they could each be used to transmit 24 digitized voice channels, provided that an investment in electronics was made at each end of the wire pair. This idea proved economical, since digital electronics, using time-division multiplexing (TDM) were cheaper than analog electronics using frequency-division multiplexing (FDM); and bandwidth limitations on a pair of copper wires over short distances did not present a problem. Also, a TDM receiver only had to distinguish between a 0 or a 1, whereas an FDM receiver would need to determine the amplitude and frequency of an incoming signal, requiring a much higher quality signal and a more expensive receiver. In addition, analog FDM allowed crosstalk (interference) to occur among adjacent voice channels; digital TDM did not. On the other hand, in typical long-haul, high-density transmission systems, especially microwave radio systems, the availability of bandwidth is usually very critical, and digital PCM neither provided sufficient economical improvements nor technical improvements over analog techniques to justify its use in long-haul systems. This is still true today, but is becoming less so as time goes on.

There are two trends that make digital appear to be the way to go in the future. The first is that more telephone company central offices are becoming digital and are switching voice traffic digitally. Thus, even if digital transmission is more expensive than analog, it may pay to use digital transmission facilities between adjacent digital central offices, since a telephone company can then save the investment in channel banks, or analog-to-digital and digital-to-analog converters, at both ends of the interconnections. Second, since digital repeaters regenerate a new signal instead of amplifying an analog signal, signal quality can be better. A digital repeater only emits 0s or 1s, so that

TABLE 2-1
Analog and Digital Signal Equivalents

Type of Analog Signal	Analog Bandwidth Requirement (BR) (Hz)	Sampling Rate PAM Pulses (2×BR) (samples/second)	Number of Quantizing Level Desired (2^n bits)	Number of Bits Needed to Produce Required Quantizing Levels	Bit Rate†
Telephone voice	4,000 Hz	8,000	128 levels = 2^7	7 bits	56,000 bps
			256 levels = 2^8	8 bits	64,000 bps
Lo-fidelity music (AM radio)	4,000 Hz	8,000	1,024 levels = 2^{10}	10 bits	80,000 bps
Hi-fidelity music (FM radio)	15,000 Hz	30,000	1,024 levels = 2^{10}	10 bits	300,000 bps
Picturephone meeting service	1,000,000 Hz	2,000,000	8 levels* = 2^3	3 bits	6,000,000 bps
Color TV	4,800,000 Hz	9,600,000	1,024 levels = 2^{10}	10 bits	96,000,000 bps
Still picture-video telephone	4,000 Hz (or higher)	8,000 or higher	8 levels = 2^3	3 bits	24,000 bps or higher
Still picture on television screen	40,000 Hz	80,000 or higher	64 levels or higher = 2^6	6 bits or more	480,000 bps or higher

Type of Digital Signal	Bit Rate
International telex	50 bps
TWX (teletypewriter exchange)	110 bps
Dataphone Digital Service (DDS), offered by AT&T	2,400; 4,800; 9,600; 56,000; 1,544,000 bps
Low-speed data	less than 1,200 bps
Medium-speed data	1,200 to 9,600 bps
High-speed data	over 9,600 bps
Videotex (Viewdata)	1,200 bps

* In reality, a wider bandwidth with data compression is used to obtain the 6 million bps rate.

† Number of PCM bits per second (bps) needed to carry analog signal (sampling rate × bit rate) $8,000 \dfrac{\text{samples}}{\text{second}} \times 7(\text{or } 8) \dfrac{\text{bits}}{\text{sample}}$.

as long as it can distinguish between a 0 or 1 incoming signal, it can produce a new 0 or 1 outgoing signal. An analog repeater only amplifies an analog signal, so it amplifies the noise as well as the original signal. Thus the signal-to-noise ratio deteriorates after each successive analog repeater. Third, the cost of digital electronics is dropping rapidly, thus making digital transmission facilities less expensive. Since digital electronic devices are generally simpler than analog devices, digital devices lend themselves more to large-scale integration (LSI) than do analog devices. Therefore unit costs of digital equipment are expected to continue to decline in cost over time. At the same time, the amount and value of installed digital hardware continues to increase as more digital central offices are installed each day. Therefore a continued transition from analog to digital transmission facilites is expected over time.

TRANSMISSION SYSTEMS

This section will discuss various types of transmission systems, both analog and digital, their historical place in the telecommunications network, and what we anticipate their future use to be.

Wire Pairs

The oldest and simplest type of transmission systems are open wire and twisted wire pairs. With open wire pairs, two uninsulated wires are strung in parallel, as on a telephone pole, and not too close together (at least 6 inches apart), to complete a communications circuit. With twisted wire pairs, adjacent pairs are insulated and can be spaced relatively close. Since wire pairs are generally bunched together to form a larger cable, they are twisted together in pairs to avoid inductive interference, which creates crosstalk among adjacent wire pairs.

Coaxial Cable

Coaxial cable was developed to counter the severe attenuation, or loss of signal strength, that occurs at higher frequencies. As the signal frequencies become higher, the electrical current tends to flow more on the outside edge of a wire. This is known as the "skin effect." The current uses an increasingly small cross section of the wire during propagation; therefore the effective resistance of the wire during higher frequency propagation increases, causing increased attenuation or signal loss. Also, at higher frequencies an increasing amount of the signal's energy is lost by radiation from the wire (the wire gives off more heat). Still, it is desirable to transmit at as many high frequencies as possible so that as many separate signals, or channels, as possible can be sent over the same wire transmission facility. However, the skin effect limits the higher frequencies.

Coaxial cable can transmit signals at higher frequencies than can wire pairs. A coaxial cable consists of a hollow copper cylinder (perhaps woven copper filaments), or other cylindrical conductor, surrounding a single copper wire conductor. The space between the cylindrical shell and the inner conductor is filled with an insulator. At higher frequencies the cable is relatively immune to noise, and there is virtually no crosstalk or interference between adjacent coaxial cables because the current tends to flow on the inside of the outer shell and the outside of the inner copper wire. Because of this shielding from noise and crosstalk, the signal can be transmitted at higher frequencies and can be dropped (or suffer degradation) to a lower level before amplification is required and still retain most or all of its information content. Thus the reason for higher capacities in coaxial cables, relative to wire pairs, is that signal loss, or attenuation, does not become severe until very high frequencies.

The first coaxial cable telephone system was installed in the Bell System in 1941. It was a relatively low density system compared with today's standards; it could only carry 600 voice channels per coaxial cable and included only four cables per sheath. Thus each sheath could handle only 2,400 one-way voice channels.

Terrestrial Microwave

The main contender with coaxial cable systems or bulk transmission is microwave radio, a technology somewhat newer than coaxial cable and developed from radio detection and ranging technology (RADAR), which was first designed by the U.S. Naval Research Laboratory in 1934 and perfected during World War II. Microwave radio was introduced into the Bell System network in the early 1950s. Unlike high-frequency or longer-wave radios that bounce signals off the ionosphere—such as AM radio or shortwave radio transmitters—a microwave radio beam employs line-of-sight transmission, and the antennas used in each microwave radio relay station are all on towers within sight of one another. Since relay towers are normally spaced every 20 to 30 miles because of the curvature of the earth, on a long-distance transmission via microwave the signal must be picked up every 20 to 30 miles and retransmitted.

One advantage microwave radio has over coaxial cable is that it requires amplifiers only every 20 to 30 miles verses every 1½ to 4 miles for coaxial cable systems. (Another advantage is that microwave does not require cables in trenches or on telephone poles in areas of rough or mountainous terrain.) The disadvantage of using too many amplifiers, as in coaxial analog-type cable systems, is that a slight defect in any one of the amplifiers is cumulative, and analog amplifiers amplify the noise as well as the signal. Thus the more amplifiers one uses, the greater the signal degradation. As a result of its characteristics, microwave radio is used to carry about 70 percent of all interstate long-distance telephone calls in what was the Bell System. There are about 410 million circuit miles in the microwave radio relay network of the former Long Lines department of AT&T (now AT&T Communications).

The analog microwave radio industry is relatively mature, and unit growth is generally less than long-distance traffic growth in terms of complete systems. The major manufacturers of analog microwave radios are Western Electric, GTE Communication Systems, Nippon Electric, the Farinon division of Harris Corporation, and the Collins Transmission Systems division of Rockwell International.

Optical Fiber

Lightwave communications date back to the 1870s. In fact, Alexander Graham Bell invented the photophone before he invented the telephone. He considered it to be his greatest invention. The photophone used sunlight to modulate an electric current (see Figure 2–12). Talking into the photophone mouthpiece would make the diaphragm of the mouthpiece oscillate at the frequency and amplitude of one's voice, the same way a speaker's voice makes the diaphragm and electric current oscillate at the frequency and amplitude of his or her voice in a normal telephone. On the opposite side of the diaphragm was a mirror. The angles and the frequency of the changes in the angles of the light waves hitting the mirror on the backside of the diaphragm would coincide with changes in the position of the diaphragm, which was vibrating due to the speaker's voice. The photocell of the receiver could pick up the intensity of the angle and frequency of the changes in the angles of the light waves hitting it. The photocell could then convert these vibrations into changes in an electric current it would produce in the receiver, which would

FIGURE 2–12
Simplified Photophone Schematic

operate a speaker and reproduce the speaker's voice. The obvious problem with this system is transmission. Nothing could interfere with the light waves between the mouthpiece and the receiver. Fog was a real problem. Thus Bell abandoned the photophone and went on to invent the electricity-powered telephone.

In the mid 1960s an ITT scientist presented a paper discussing the practicality of lightwave communications. To solve Alexander Graham Bell's problem of fog and other forms of atmospheric interference, he theorized that lightwave signals could be transmitted over fibers of ultrapure glass. Optical fiber technology was originally expected to become of commercial significance after the turn of this century. The telephone industry expected microwave waveguide pipes to become the next major advance in transmission technology. A microwave waveguide pipe is a hollow tube through which microwave radio signals are transmitted. It functions like a microwave radio whose path is guided by a metallic tube. Because the radio path is shielded from weather and other obstacles, waveguide pipes can operate at higher frequencies and can carry more traffic than can normal microwave radios and without running into the fading and other reliability problems microwave radios generally experience. However, microwave waveguide pipes never really got very far past the experimental stage due to the quicker-than-expected development of optical fiber systems technology. By the mid-1970s, it was obvious that optical fiber systems technology had arrived, and experimental systems were built and tested by AT&T, General Telephone & Electronics, and others. By 1980, optical fiber systems were becoming cost competitive with coaxial cable systems and, because of their potential capacity, they were beginning to be installed not as experimental systems but as regular commercial equipment.

Optical fiber systems in concept are similar to coaxial cable systems except that optical fiber transmits light and coaxial cable transmits microwaves. Light can be considered to be electromagnetic radiation with very high frequencies and very short wavelengths, while microwaves are somewhat lower in frequencies and longer in wavelengths. Optical fiber systems are essentially made up of three principal elements: (1) a source or a transmitter, which converts electrical energy into light energy; (2) an optical fiber cable, or optical waveguide, which is used to carry the light signals from the transmitter to the receiver; and (3) a receiver or detector, which reconverts the light energy, or photons, back into electrical energy.

Optical transmitters. There are two basic types of lightwave communications—guided and unguided. Guided systems employ optical fiber cables (hair-thin fibers of transparent glass or plastic) as a conduit to guide the light beams from the source to the detector. Unguided systems do not utilize fiber optic waveguides but instead rely on a direct beam of light from the source to the detector. Because they have problems similar to the photophone, we do not see many applications of unguided optical systems, and thus we have concentrated our discussion on guided optical systems.

In a guided optical fiber system, the transmitter converts an electrical signal into light energy. In digital systems, the light energy is modulated into bursts of light that represent an information code. For example, in the code utilizing the binary number system, the basis of all digital communications (recall pulse code modulation), a burst of light could represent a 1, and the absence of a burst of light could represent a 0. (Obviously we are discussing a clocked or synchronous operation.) Then the modulated light energy is transmitted (guided) through an optical fiber to the receiver (detector). At the receiver the light energy is reconverted into an electrical signal and decoded. The signal then continues to be transmitted as a flow of electrons or radio waves. An additional component, called a repeater, is used on long-distance links in order to amplify and reshape photon streams weakened by attenuation, or loss of signal strength, over a certain distance.

There are several parts to the transmitter device. It is designed to encode the signal, drive the light source, emit the photons, and couple the emitter to the fiber. The primary electro-optical component is the emitter. There are two general types of emitters (light sources): (1) incoherent (out of phase) light-emitting diodes (LEDs) and (2) solid-state injection laser diodes. (A diode is a two-electrode semiconductor device that transmits current more easily in one, and usually only one, direction. *Laser* stands for light amplification by stimulated emission of radiation.) Both of these devices can be modulated at very high speeds. For example, the transmission bandwidth for infrared LEDs can be as high as 150 MHz and that of solid-state lasers as much as a few GHz. In comparison, the transmission bandwidth for most telephonic coaxial systems is less than 100 MHz, and in some systems it is much less.

In terms of the theoretical potential, LEDs are inferior to injection laser diodes because they cannot concentrate as much light into the fiber and because they cannot be modulated as rapidly. Still infrared LEDs do meet the requirements for many optical fiber transmission applications today. Over the longer term though, the higher potential capacity of solid-state injection laser diodes, combined with their declining cost and increasing reliability, could result in their utilization in most future optical fiber systems.

Types of optical fibers. The optical fiber cable is composed of one or more optical fiber waveguides. The individual fibers are made in hair-thin strands in order to provide a flexible cable. There are two basic types of optical cable: bundled-fiber cable and isolated-fiber cable. The bundled-fiber cable is composed of a group of individual fibers bundled together and sharing a common source of light. This type of cable is normally used in short-distance applications and is well suited for LEDs because the bundle provides an aperture large enough to collect the noncoherent (out of phase) light of the LED. The isolated-fiber cable is also composed of a number of individual fibers. However, each fiber transmits its own signal, different from that of the fiber next to it, because each fiber uses its own separate light source. Isolated-fiber cables are generally used with injection laser diodes in both short- and long-distance applications.

Of the types of optical fiber waveguides in use today, the two most important are the stepped-index fiber and the graded-index fiber. Figure 2–13 illustrates how these two types of fibers guide the waves of light through the emitter. The stepped-index fiber, usually made from plastic, is composed of a central core and an outer cladding. Since the index of refraction (or reflection) of the core is higher than that of the cladding, light striking the core-cladding interface is reflected into the core. (The core-cladding interface acts like a mirror.) Thus light is propagated through the fiber following the path of the cable. Graded-index fiber, usually made of fused silica (glass), uses a gradual variation in the refractive (reflective) index that forces the light rays to propagate along the fiber in a wavelike manner. This wavelike movement lessens pulse broadening, which can distort signals over a long distance. Thus the stepped-index fiber is generally used in shorter-distance applications, and the graded-index fiber is used mainly for longer-distance communications.

Short-distance optical fibers systems can use plastic instead of glass for the composition of the optical fiber cable. Advances are being made with plastic optical fibers as well as with glass ones. In fact, Nippon Telegraph

FIGURE 2–13
Optical Fiber Waveguides

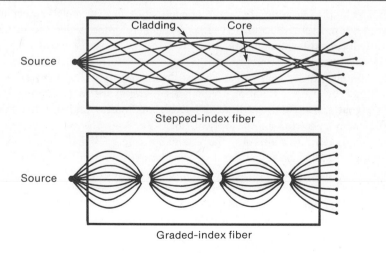

Stepped-index fiber

Graded-index fiber

and Telephone Public Corporation (NTT), Japan's national telephone company, has developed a new plastic optical fiber that will extend short-distance light communications by some 30 times over current levels. The new fiber reduces the loss of light passing through it to one third of the loss of today's best fibers. This new fiber transmits as far as up to 300 meters. Present conventional plastic fibers do not permit light signal transmission of more than 10 meters in length.

The fiber, which is made of an acrylic resin, has the added features of being less costly to produce and more pliable to work with than its glass fiber counterparts. NTT envisions using the fiber for intrabuilding and factory communications systems. NTT expects to have the fiber ready for practical application within the next year or two.

Characteristics of optical fiber. An important characteristic of optical fibers is their loss factor and attenuation characteristic. Losses in the fiber result in attenuation or weakening of the light signal and limit the length over which a fiber optic transmission system can operate without the need for signal repeaters. Losses are caused primarily by light absorption and scattering. Absorption losses are caused by impurities present in the fiber. The most significant loss factor is due to the scattering or dispersion of light by the molecular structure of the fiber. This type of loss can be minimized by optimizing the relationship between the characteristic wavelengths of the emitter and the molecular structure of the fiber.

Since glass is noninductive, fiber optic transmission is immune to lightning, inductive crosstalk, electromagnetic effects induced by such things as nuclear explosions, and generally all types of conventional electromagnetic interference. This characteristic is one of the two most important advantages of optical fiber transmission. (The second is the large available bandwidth.) For example, in computer interconnections it is extremely important that data be transmitted accurately. Electromagnetic interference can distort signals during transmission and thus lead to the transmission of data with a significant number of errors.

Another advantage of fiber optic transmission is its inherent large bandwidth. The wave oscillation frequencies of light are so rapid that the rate of signal transmission is limited by the modulator, and not the available frequency spectrum in the optical fiber. Thus, as the trend toward faster modulation rates continues, the attainable rate of signal transmission will rise. The bandwidth advantage is of considerable importance to many potential users of fiber optics, particularly the telephone industry. For example, Bell Labs has indicated that a single optical fiber as thin as a human hair can transmit as many phone conversations as a bundle of copper wires as thick as a man's arm. As a result of the bandwidth advantage, optical fibers are very small and lightweight, in comparison with the bundles of copper wire required at equivalent transmission rates. Also, as faster modulation devices are invented, the electronics of an optical fiber system can be replaced to increase transmission speeds without "digging up the streets" to install a higher capacity optical fiber, since the optical fiber already has greater bandwidth capabilities than are being utilized in today's optical fiber systems.

Optical fibers are also potentially very inexpensive, since glass is a common substance. The raw materials for glass are as abundant as sand (silicon dioxide), whereas copper is in increasingly short supply. This economic incentive could be particularly important to the telecommunications industry. For example, we believe that AT&T and its formerly affiliated companies account for almost one fifth of domestic copper consumption. During the next couple of years, fiber optic networks may be priced such that the newer technology will be less than half the cost of the conventional technologies.

It is also important to note that optical fibers are generally unaffected by moisture and temperature so that cable pressurization is not required. For example, underground telephone cable is jacketed and pressurized in order to protect the wires from moisture. The maintenance of pressure throughout the cable can be an expensive process. Also, fiber optic transmission presents no short-circuit hazard as do copper wires and cable. This can be important for control applications in potentially hazardous environments, such as a control system at a chemical refinery or an electric utility power plant. In fact, electric utilities have already begun to use optical fiber systems for communications purposes at power plant facilities.

Historically, one of the most important factors impeding the progress of fiber optics was the high unit costs, which stem from insufficient production volume for the component suppliers. The technology is in place today to design affordable optical fiber links, but the current lack of commercial volume tends to maintain a relatively high unit cost on the optical components, although unit prices have been dropping over time.

Another problem had been field maintenance and installation. Connecting and terminating an optical fiber involves significantly different techniques than conventional wiring. However, just about all of the problems involving the installation and the splicing of optical fibers have been solved.

The telephone industry represents the largest near-term potential market for fiber optic cable. (Longer term, the cable TV industry could employ optical fiber systems to distribute television signals.) It is believed that the domestic telephone industry installs about 200 million feet of coaxial cable annually. In addition, its total U.S. consumption of noncoaxial exchange wire is about 400 billion conductor feet. The initial application of fiber optics in the telephone industry has been short-distance central exchange interconnection, via the underground conduits in large cities. Presently the conduits in metropolitan areas (such as the Chicago site of an AT&T optical fiber experiment) are overcrowded and significantly limit the telephone industry's ability to handle increased telephone calling volumes. Since fiber optic cable can displace large bundles of copper wire pairs, this technology can alleviate that situation.

The emergence and acceptance of optical fiber transmission equipment for digital trunking applications and the fact that several systems are in service today have increased the options for transmitting information. Comparison of fiber optics with conventional cable or microwave radio technology is usually required before a transmission system is selected.

Communications Satellite Systems

The concept of a communication satellite is relatively simple. It is, in essence, a microwave radio relay station in the sky. Signals are sent to it from ground stations, are amplified within the satellite, and are sent back to earth. The value of a communications satellite lies in the fact that it can handle a large volume of traffic and can send it almost anywhere on earth. The cost of satellite channels is dropping rapidly, and their capabilities are improving.

Like a microwave radio relay station, a satellite receives a radio signal at a certain frequency, changes its frequency, amplifies it, and then retransmits it. The equipment that performs this function is called a transponder. The radio signals have a high bandwidth (typically 36 MHz), which is important because the information-carrying capacity of a signal is proportional to its bandwidth. Most commercial communications satellites use the same radio frequencies (RF) as terrestrial microwave relay systems. The main difference between microwave radios and satellites is that microwave radio stations are typically spaced 20 to 30 miles apart, and satellites are 22,300 miles up in space.

A significant problem with satellite communications is the delay that occurs between a signal's transmission and its reception. A one-way transmission takes about 270 milliseconds (thousandths of a second), so that a round-trip communication, like a conversation, takes about 540 milliseconds (more than half of a second) to complete (see Figure 2–14). This results in the talker's voice being returned as an echo. This problem can be overcome through the use of echo suppressors (long used on any land circuit exceeding 1,500 miles) and the newer echo-cancellation devices. The delay particularly affects data streams. As a result, a technique called forward error correction (FEC) has replaced the older "wait for acknowledgement" form of digital transmission.

Satellite frequency bands. Like a terrestrial microwave radio system, a satellite uses different frequencies for the reception and transmission of signals. Without this arrangement, a powerful transmitted signal would interfere with a weak incoming signal. Table 2–2 lists the radio-frequency (RF)

FIGURE 2–14
Travel Time in a Satellite System

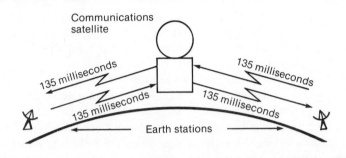

bands designated for all forms of radio communications. Most communications satellites use the UHF and SHF frequency bands; commercial communications satellites use the SHF band exclusively and will continue to do so for the foreseeable future.

Frequency bands are also given letter designations. Commercial communications satellites use three RF bands, the C band (4 and 6 GHz), the Ku band (11 or 12 and 14 GHz), and the Ka band (20 and 30 GHz). The lower frequency of each band is used for the downlink; the higher frequency for the uplink. Many designations follow the signal path. Thus 14/12 or 6/4 GHz is often found.

TABLE 2–2
Radio Frequency (RF) Bands

Band Number	Band Name	Frequency Range		Wavelength
4	VLF—very low frequency	3–30	kHz	Myriameters (10^4m)
5	LF—low frequency	30–30	kHz	Kilometers (10^3m)
6	MF—medium frequency	300–3,000	kHz	Hectometers (10^2m)
7	HF—high frequency	3–30	MHz	Decameters (10 m)
8	VHF—very high frequency	30–300	MHz	Meters
9	UHF—ultra high frequency	300–3,000	MHz	Decimeters (10^{-1}m)
10	SHF—super high frequency	3–30	GHz	Centimeters (10^{-2} m)
11	EHF—extra high frequency	30–300	GHz	Millimeters (10^{-3} m)
12		300–3,000	GHz	Decimillimeters (10^{-4} m)

Most commercial satellites use the C band (4/6 GHz). However, since these are the same frequencies used by terrestrial microwave radio systems, newer satellites are beginning to use the Ku band. The Ka band will probably not be used, except for experimental purposes, until the 1990s.

Table 2–3 gives a comparison of the radio frequencies allocated by the FCC for terrestrial microwave and satellite communications.

Some of the newer satellite systems will operate in the Ku band. As explained by James Martin, the advantages of the Ku band are:

1. The band is currently less heavily used for terrestrial common carrier microwave radio relay stations, so 12/14 GHz earth antennas can operate on the rooftops of buildings in cities. Large corporate locations can operate their own antennas. In congested areas there may be one earth station serving many local users who are linked by short line-of-sight packet microwave radio systems, such as those proposed by M/A-COM and others 10.55–10.68 GHz (RAPAC—radio packet), or coaxial cable (CAPAC—cable packet) networks, possibly supplied by the cable TV companies. In fact Local Digital Distribution Company (LDD), a subsidiary of M/A-COM, plans to offer such services to be used with the Satellite Business Systems network.

TABLE 2–3
Common Carrier Radio Frequencies

| Band | Terrestrial Frequency Bands (GHz) | Commercial Satellite Frequency Bands | | Bandwidth (MHz) |
		Downlink (GHz)	Uplink (GHz)	
S band	2.11–2.13			20
	2.16–2.18			20
C band	3.7–4.2	3.7–4.2		500
	5.925–6.425		5.925–6.425	500
Ku band	10.7–11.7			1,000
(frequencies of		10.95–11.2		
the new		11.45–11.7		500
generation of		11.7–12.7		500
satellites)			14.0–14.5	500
Ka band	17.7–19.7			2,000
		17.7–20.2		2,500
			27.5–30.0	2,500

2. The beamwidth from a 12/14-GHz earth station antenna of a given size is less than half of that for a 4/6 GHz satellite. It is inversely proportional to the frequency. Therefore, many more Ku band satellites could be operated without any increased signal interference among adjacent satellites, thereby reducing potential congestion in the equatorial orbit.

3. A satellite antenna of a given size and weight can be made more directional by using a higher frequency. Multiple-spot (searchlight) beams to or from the satellite could therefore be made to operate at the same frequency. The satellite could then transmit more signals than a 4/6-GHz satellite without exceeding the 500 MHz radio frequency bandwidth allocation. Also, when using small three-meter earth-station antennas, Ku band satellites can be spaced at 2 degrees apart in orbit and still be able to transmit an intelligible signal, instead of the 3 degrees that, may be required to transmit to three-meter earth-station antennas from satellites that operate in the C band.

4. When the 4 GHz downlink is used, there is a lower power limitation imposed on the radiated power of the satellite to prevent interference with terrestrial common carrier systems than is imposed on the 12 GHz downlink. Again, the higher the radiated power of the satellite, the smaller the earth-station antenna required to receive and be able to understand the signal.

   Note, though, that the higher frequencies also have some disadvantages. With very heavy rain, the received signal's strength falls, and the noise the signal picks up increases. Also, most 12/14 GHz links need to be designed to avoid low angles of earth-station antenna elevation where the path through any rain is long.[1]

Satellite orbits. What makes a satellite stay aloft? A satellite stays in orbit because centrifugal force caused by a satellite's rotation around the earth exactly balances the earth's gravitational pull. The satellite's velocity would pull it away from the earth if gravity did not exist. However, the earth's gravity exactly balances the effect of the satellite's velocity.

The closer a satellite is to the earth, the stronger is the earth's gravitational field, and the faster a satellite must revolve around the earth to avoid falling to earth. Low-orbit satellites travel at about 17,500 miles per hour and revolve around the earth in about one and a half hours. Communications satellites travel at 6,900 miles per hour and revolve about the earth in the period of earth's own rotation, 23 hours, 59 minutes, and 4 seconds.

A low-orbit satellite revolves about the earth at a height of from 100 to 300 miles and is in the line of sight of an earth station about 15 minutes or less. A medium-orbit satellite revolves at from 6,000 to 12,000 miles and is in sight of an earth station from 2 to 4 hours. A geosynchronous satellite, one that revolves about the earth in a sidereal day, revolves at an altitude of approximately 22,300 miles. Since a geostationary satellite revolves around the earth about every 24 hours, the period of an earth day, it appears stationary to the observer on the ground, or geostationary. It is always in a consistent line of sight of an earth station looking at it. Geostationary satellites orbit the earth above the equator.

The geostationary orbit offers many advantages in satellite systems engineering:

1. The satellite appears almost stationary in orbit relative to the earth-station antennas, so that the cost of computer-controlled tracking of the satellite is avoided. A nonmovable antenna can be used that may need only minor occasional manual adjustments.
2. There is no necessity to switch or transfer from one communications satellite to another as one disappears over the horizon and another one appears.
3. There are no regular interruptions in transmission. A geostationary satellite is permanently in the view of an earth station communicating through it.
4. Because of its distance above the earth, a geostationary satellite is in the line of sight of more than 40 percent of the earth's surface (38 percent if angles of elevation below 5 degrees are not used). The areas not covered are the polar regions and the other sides of the earth. A large number of earth stations can see, or look at, the same satellite and therefore can intercommunicate.
5. Three satellites can give total global coverage with the exception of the North and South Poles. The INTELSAT system uses three groups of satellites (in the Atlantic, Indian, and Pacific Ocean regions) to provide coverage of most of the world's population.
6. For most applications there is no noticeable Doppler shift in the radio signal, that is, the change in the apparent frequency of the signals going to and from the satellite. The Doppler shift is caused by the motion of the moving satellite as it approaches and passes the earth station. Satellites in elliptical orbits have different Doppler shifts for different earth stations, and this increases the complexity of the receivers, especially when large numbers of earth stations intercommunicate. Geostationary satellites do not appear to move relative to the earth station. Thus no (or a very small) Doppler shift arises.

Disadvantages of geostationary satellites include these:

1. Latitudes greater than 81.25° north and south (or 77° if angles of elevation below 5° are excluded) are not within sight of the satellite's radiation at the longitude of the satellite (lower for other longitudes). Fortunately, there is not much more than polar ice at these latitudes.
2. Because of the altitude of the satellite, the received signal power, which is inversely proportional to the square of the distance and frequency

between the earth and the satellite, is extremely weak. Also, the signal propagation delay is about 270 milliseconds for the combined uplink and downlink in one direction.

3. There are a limited number of orbital slots available over the equator in any given frequency band. Geostationary satellites using the same radio frequencies and covering the same areas of the earth cannot be placed too close together because their signals will interfere with each other. Today, C-band satellites are placed 4 degrees apart in orbit. The FCC has recently approved 2-degree spacing, although this would require using larger earth-station antennas. Satellites may be colocated if they use different frequencies or serve different nations with nonoverlapping beams—North and South American countries, for example.[2]

Satellite in-orbit spacing depends upon many factors, including the design of both the space segment and the ground segment. Spacing is affected by the beamwidth of the transmitting earth station. The beamwidth varies with the size of the antenna's aperture and the frequency band used. The width of a beam from the satellite is inversely proportional to the width of the satellite's transmitting antenna. The same holds true for the earth station. The wider the antenna, the narrower the beam (and vice versa). Reason: A larger antenna can more precisely focus a beam, like a searchlight. One might think that the trend toward smaller earth-station antennas is contrary to narrower spacing between adjacent satellites. However, the use of higher frequencies (Ku and, eventually, Ka bands) will actually permit the operation of more satellites for two reasons: (1) A new radio frequency (RF) band will not interfere with existing ones, and (2) the beamwidth is inversely proportional to the frequency. The higher the frequency, the narrower the radio beam for a given antenna size.

The uplink is generally operated at the higher frequency, in any band. The downlink is operated at the lower frequency, since the higher frequency emits a narrower beam radiating into space. Use of the lower frequency to downlink is important since satellites have limited antenna diameters and power for the transmitters, but earth stations do not. The use of a lower frequency in the space-to-earth direction means lower path losses and allows the use of a lower transmitter power in the satellite.

The usable equitorial orbital arc that exists over the United States runs from 58° west longitude to 143° west longitude, or an arc of 85 degrees. At 4-degree spacing, about 21 satellites in a particular RF band can be accommodated; at 3-degree spacing, about 28 satellites. However, some orbit slots must be set aside for Canada, Mexico, and so on.

DATA TRANSMISSION CODES

Thus far we have discussed encoding and modulation techniques for analog communications. Digital communications also employ codes. These codes are used in the transmission of characters and numerals.

Baudot

Baudot is the simplest and slowest of all the major codes in use today. Baudot is a 5-bit code that yields 32 combinations or possible characters ($2^5 = 32$). The first 26 combinations are used for the letters of the English alphabet. The 27th and 28th combinations are used as control characters. The 29th combination designates a downshift to lowercase, which indicates that 5-bit codes 1 through 26 represent letters. The 30th combination designates an upshift to uppercase, which indicates 5-bit codes 1 through 26 represent either number or punctuation symbols. The 31st combination represents a space, and the 32d combination is not used.

The Baudot code is used for national and international Telex communications, such as those domestically provided by Western Union Telegraph, and the international record carriers, such as ITT World Communications and RCA Global Communications. Baudot is also used in most teletypewriter devices (TTDs) for the deaf.

The Baudot code is usually transmitted at 50 bits per second, or 66.67 words per minute, or about 6.7 characters a second, which equals 7.5 bits per character. Five bits per character are used for the code, and 2.5 bits per character are used for control purposes. Of the 2.5 bits used for control, 1 bit signals the start of a character, and 1.5 bits signals the end of a character.

ASCII

The ASCII code (pronounced askee) is a seven-bit code (ASCII stands for American Standard Code for Information Interchange). With seven coding bits, 128 combinations are possible ($2^7 = 128$). However, ASCII is an eight-bit coding system. It uses a seven-bit code to encode the character plus an eighth bit for a parity check. In computer terminology, each eight bits is called a *byte*. A parity check is the addition of noninformation bits to data, making the number of ones (1s) in a grouping of bits either always even or odd. This permits detection of bit groupings that contain single errors.

In such public communications systems as Western Union Telegraph's TWX (teletypewriter exchange) service, data transmission generally occurs at 110 bits per second or 100 words per minute. (There are five characters in the average teletype word. With one character for spacing, the average word is six characters. In general, x words per minute equal $x/10$ characters per second.) The 100 bps works out to 10 characters a second or 11 bits per character. Seven bits encode a character, one bit is used for parity and error checking, and three bits for control (start-stop) purposes. One bit signals the start of a character, and two bits signal the end of a character. Don't forget this is asynchronous transmission and start-of-character and end-of-character signaling is required.

EBCDIC

A third encoding scheme is called EBCDIC (extended binary coded decimal interchange code). EBCDIC is an eight-bit coding system, was developed by IBM for use on IBM computers, and is an example of a proprietary encoding scheme, although widely used. It allows up to 256 combinations ($2^8 = 256$). EBCDIC has been used for a long time in computers. Whereas Baudot and ASCII are used principally in teletypewriter operations, EBCDIC is used principally with business computers and in computer-to-computer or computer-to-data terminal types of systems. EBCDIC uses eight bits (256 combinations) to operate printing devices with up to 256 graphic characters or to transmit eight-bit bytes of computer data. However, the ASCII code is widely used in microcomputers and is the de facto standard for microcomputer communication and public and consumer data bases and services such as Compuserve, Dialog, and so forth.

MODES OF TRANSMISSION

There are three basic modes of transmission in data communications: simplex, half duplex, and full duplex. With simplex transmission, data can be sent from point A to point B but not vice versa. The direction of data transmission can be in one direction only, much like a commercial radio station can broadcast music but not receive it.

Half Duplex

Half duplex allows transmission in either direction but not in both simultaneously. Half duplex is like citizens band (CB) radio. One person talks and then listens. One cannot talk and listen at the same time on CB radio because as soon as the microphone is turned on, the speaker cuts out. Releasing the microphone button returns the speaker to operation. Half duplex is generally used in data communications systems where only a two-wire circuit is available. The same circuit is used for both directions of communications, and thus the entire bandwidth of the channel is devoted to the transmitter by only allowing one direction of traffic at a time. The other side can only transmit when the first side is idle.

Full Duplex

Full duplex allows transmission in both directions at once. Full duplex usually requires a four-wire circuit, a complete circuit for each direction of transmission. Full duplex can be performed over a two-wire circuit by dividing the circuit into two separate frequency bands. (Obviously neither direction of transmission can use the full available bandwidth of the channel if it has to share it, as it does in full duplex.) One frequency band is used for transmission in one direction and the other band for the opposite direction. Keeping two signals separated in frequency prevents them from interfering with each other. Also both directions need not have the same bandwidth or bit speed. Some videotex systems, such as Viewdata, receive data at 1,200 bps but transmit at 75 bps. However, full-duplex transmission over a two-wire circuit limits the maximum speed in any single direction to something less than the maximum speed capable of the channel if half-duplex transmission were used, due to the sharing of the channel and the guard band needed between each direction of signal transmission.

Asynchronous

Another aspect of data communications is synchronous and asynchronous transmission. Asynchronous transmission is often called start-stop. A single character is sent at a time. As with Baudot and ASCII codes, each character begins with a start signal, or bit(s), and ends with a stop signal, or bit(s). With start-stop transmission, there can be a varying interval between one character and the next. When one character ends, the receiving terminal sits and waits for the start of the next character. The transmitter and the receiver begin operation together with the start bit of a character, they remain synchronized while the character is being sent, and end operation together with the stop bit(s) of the character. Asynchronous transmission is used in most teletypewriter applications.

Synchronous

Synchronous transmission is used when the communicating devices transmit to each other with regular timing. When devices transmit to each other continuously, synchronous transmission can give the most effective line utilization. With synchronous transmission, the bits of one character are immediately followed by the bits of the next character. There are no pauses and no start or stop bits between characters. Groups of characters are divided into blocks, which are sent down the communications pipeline at equal time intervals. The transmitting and receiving terminals have to be exactly synchronized for the duration of each block, so that the receiving device knows which is the first bit of each character, and thus knows which are the other bits of each character. Synchronous transmission is normally used in applications where a computer is at either or at both ends of the transmission facility (or line).

MODEMS

In many environments, the terminals and computers are separated by several hundred feet or, in some cases, more than a mile. For these short distances,

FIGURE 2–15
Modem or Data Set

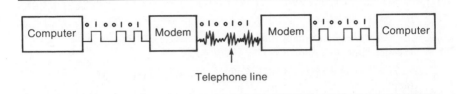

Telephone line

a direct wire connection is possible using copper wire pairs or coaxial cable. Signals are transmitted digitally, eliminating the need for modems or acoustic couplers. In terms of expense, hardwiring is usually the lowest-cost data transmission line available. Digital signals cannot travel more than 50 or so feet over wires or cables before they become unintelligible. However, they can travel thousands of miles when converted to the analog mode by an acoustic coupler or modem and sent over telephone lines.

An acoustic coupler converts a serial bit stream of data from a terminal or computer into audio tones. It then sends the tones over the phone line via a telephone handset. At the destination, the audible tones are reconverted into the original serial-bit stream of data. Acoustic couplers are used in applications with terminals that operate at 1,200 bps or lower.

Whereas an acoustic coupler converts digital data into audible tones, a modem also converts the data directly into electrical signals. As we said, telephone companies use the term *data set* to describe this device. (In fact the word *modem* is rarely, if ever, used in telco marketing brochures.) Modem is a contraction for modulator-demodulator. Modems are also called data sets by the Bell System, and their application is to prepare computer or data signals for transmission over a telephone line. A data set, or data modem, prepares a digital data signal for transmission over an analog telephone line. Figure 2–15 gives an illustration.

Modems connect the digital world of data communications to the analog world of the telephone line. If one compared the data signal spectrum to the band of a standard voice-frequency channel, it would show that it is impossible to directly transmit digital signals through an analog channel. To get around this problem, a modem translates the frequency band of the digital signal to a band that can be transmitted through the analog channel.

■   **NOTES**   ■

1. James Martin, *Communications Satellite Systems* (Englewood Cliffs, N.J.: Prentice-Hall, 1978), pp. 144–45.

2. Ibid.

■ REFERENCES ■

Cohen, L. G. "Testing Today the Fibers of Tomorrow." *Bell Laboratories Record*, February 1981, pp. 49–54.

Martin, J. *Introduction to Teleprocessing*. Englewood Cliffs, N.J.: Prentice-Hall, 1972, pp. 61–70, 80–85.

_____. *Telecommunications and the Computer*. 2d ed. Englewood Cliffs, N.J.: Prentice-Hall, 1976, pp. 263–77.

_____. *Future Developments in Telecommunications*. 2d ed. Englewood Cliffs, N.J.: Prentice-Hall, 1977, pp. 193–203.

_____. *Communications Satellite Systems*. Englewood Cliffs, N.J.: Prentice-Hall, 1978, pp. 45 and 144–45.

Authors

Glenn R. Pafumi
First Vice President, Research
Dean Witter Reynolds
New York, New York

Glenn R. Pafumi's primary responsibilities for Dean Witter Reynolds include investment research of the telecommunications industries, both the common carriers and the manufacturers. Previous positions include vice president and group manager, Telecommunications Planning, for Merrill Lynch & Co., and vice president, Securities Research Division, for Merrill Lynch, Pierce, Fenner & Smith, Inc., its brokerage subsidiary. Pafumi has a B.S. from the University of California at Berkeley and M.S. and M.B.A. degrees from New York University. He is a chartered financial analyst (CFA) and a member of the Financial Analysts Federation, the New York Society of Security Analysts, the Communications Technology Analysts Association, the American Institute of Aeronautics and Astronautics, and the Institute of Electrical and Electronics Engineers.

Dr. Stanley M. Welland
Vice President and Group Manager, Technology Planning
Merrill Lynch & Co.
Bridgeport, Connecticut

Stanley Welland is accountable for the establishment and direction of a technology planning group that provides technical planning, guidance, and assistance to corporate systems and management to maintain consistency of planning and implementation with the overall objectives of Merrill Lynch. The group constantly reviews and assesses the industry developments for potential impact on the company's information processing systems. Welland works with Merrill's venture capital organization and participates in the identification and recommendation of investments in technology that will not only yield capital growth but will also provide the firm with technological leadership in providing financial services. This covers hardware, software, and services—including satellite, cable, and related technologies. Welland joined Merrill Lynch in March 1981, when he initiated Merrill's development of The Teleport project, which he managed until recently.

For 11 years Welland held assignments in the following Exxon organizations: Exxon Office Systems Company; The Mathematics, Computers, and Systems Department; the Office of Management Information Coordination; the Headquarter's Administrative Services Department; and Exxon Enterprises (Information Systems Division). His work experience encompassed the management of a broad range of projects involving the solution of technical and business problems encountered in many of the functional areas of the Exxon Corporation and its worldwide affiliates. He was the Qyx product manager assigned to Exxon Office Systems Company and a senior adviser for the Exxon Enterprises Information Systems Division, where he was responsible for developing, evaluating, and recommending strategic and tactical plans for the growth of Exxon Enterprises' business activities in the communications systems area.

He received a Doctor's degree in Engineering Science from Newark College of Engineering in 1970 and an M.B.A. in finance and business systems at Rutgers Graduate School of Business Administration, 1975.

Chapter **3**

Satellite Regulation—A Short History

Carl J. Cangelosi, Esq.
RCA American Communications, Inc.

Outline

Measured by the history of telecommunications, the domestic satellite industry is still in its infancy, with the first domestic satellites being launched in 1974. Yet the satellite business has been marked by change during this entire period. It was born with the promise of regulatory flexibility and has continually tested its regulatory barriers. Today, the marketplace has become a significant regulator of satellite communications. This chapter will trace that development.

This is a brief history. It is not complete, since too much has happened even during the 10 to 15 years of the satellite industry's short life. There are matters of historical interest that are not mentioned, such as the initial restriction on AT&T preventing it from providing specialized services over its satellites and the integration of rates to Hawaii and Alaska. Although these were important, or at least seemed so at the time, they probably have lost their significance to present readers. Similarly, this discussion is not intended for the legal practitioner looking for the definitive work on current satellite regulation.

This history is intended to give the flavor of satellite regulation. To do this, three general subjects have been chosen—satellites and earth stations because they are the basic components of a satellite system, and rates because they determine the economic viability of satellite systems.

THE EARLY YEARS, 1970–1978—FLEXIBLE BUT TRADITIONAL REGULATION

Qualified "Open Entry"

The first commercial satellite was placed in orbit in 1965 to provide international service. By 1970 there were five international satellites but none providing service to the domestic U.S. market. In that year the Federal Communications Commission (FCC) issued the first of a series of *Reports and Orders* dealing with the legal, technical, and policy issues concerning domestic satellites.[1] The FCC also invited the submission of applications for domestic satellite systems. Eight satellite system applications were filed by such companies as RCA, AT&T, and Comsat. In 1972 the FCC asked for comments on a proposed method of handling the applications.[2]

Three possibilities for satellite authorizations were considered:

One-space segment—either by selection or a consortium.

Multiple-space segments—selection on the basis of the comparative merits of the proposed systems and other public interest considerations.

Multiple-space segments— "Open entry" for all except those disqualified on policy grounds.

After receipt of comments, a *Second Report and Order* setting out the basic policies for the domestic satellite field was adopted, which enumerated the following objectives:

Maximize the opportunities for the early acquisition of technical, operational, and marketing data and experience in the use of this technology.

Afford a reasonable opportunity for multiple entities to demonstrate how operational and economic characteristics peculiar to the satellite technology can be used to provide existing and new specialized services more economically and efficiently than can be done by terrestrial facilities.

Facilitate the efficient development of this new resource by removing or neutralizing existing institutional restraints or inhibitions.

Retain leeway and flexibility in policy making so as to make adjustments as future experience and circumstances may dictate.

With this regulatory premise, the FCC found that the public interest would best be served by providing a reasonable opportunity for entry by qualified applicants. This policy quickly became known as the "Open Skies" approach, but the FCC stated it differently. It noted that the Communications Act of 1934 required the FCC to determine whether or not an applicant was financially, technically, and otherwise qualified and whether or not the proposal would serve the public interest. All applicants who qualified would be authorized. Any attempt to select or prescribe a single system or to choose one or more systems through a comparative hearing process was rejected.

Orbital Assignments

By 1973, when the FCC authorized the first domestic satellite systems, two of the original eight applicants had withdrawn.[3] This number decreased further over the next several years so only three systems were launched, as shown on Figure 3–1 on p. 68.

The launch date of each satellite is shown in parentheses. The Satcom satellites are those of RCA American Communications, Inc.; Westar, those of The Western Union Telegraph Company; and Comstar, those of Comsat General Corporation, which were built for and leased by AT&T.

At least one orbital location for each satellite company was in the orbital arc between 119° and 132° West Longitude. This area has been called the prime arc, since it permits excellent service to all of the United States including Alaska and Hawaii.[4]

FIGURE 3–1
1978—Operational Satellites

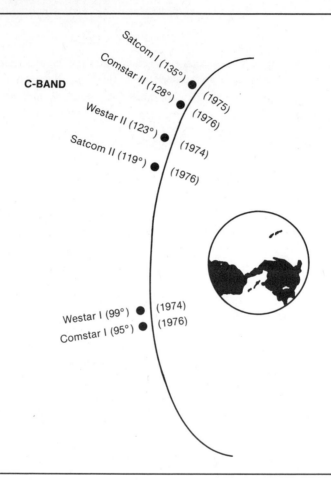

Earth Stations

In 1972 the FCC stated as its broad policy objective a flexible environment that would permit a variety of earth station ownership patterns. It favored allowing special purpose users the option of owning receive-only earth stations and did not foreclose the possibility that transmit-receive earth stations could also be owned by users in appropriate circumstances. The FCC said it would retain its intended flexibility by making determinations regarding ownership only in the context of concrete applications for particular earth stations.

Implementation of the FCC's policies started quickly. In 1973 TelePrompter Corporation requested a ruling from the FCC as to whether it needed authority under the Communications Act of 1934 to procure a transportable receive-only earth station to be used at a National Cable Television Association Convention. The FCC responded by stating that such authority was needed but granted a waiver of the requirement pending formal action.

In acting on this and similar applications, the FCC followed a three-step process: frequency coordination, construction permit, and license. Frequency coordination is the process used to resolve potential interference from other sources, such as terrestrial microwave systems. Construction permits and licenses are issued pursuant to provisions of the Communications Act granting the FCC general authority over the use of radio communication.

This three-step process was workable in the early days of the industry. The first earth stations were constructed by the satellite carriers building, launching, and operating the satellites. These stations cost millions of dollars, and the regulatory time involved in filing and obtaining the necessary authorities was not disproportionate to the investment being made. In addition, the number of earth stations was relatively small.

This situation changed dramatically from 1975 to 1977. In 1975 Home Box Office transmitted its first program to cable operators over an RCA satellite. In authorizing Florida Cablevision to construct an earth station to receive this programming, the FCC reiterated its flexible ground policy and affirmed that the proposed earth station would further its objective to afford domestic satellite applicants a reasonable opportunity to demonstrate how any operational and economic characteristics of satellites could be used to provide new specialized services.[5]

The rapid growth in the CATV industry soon forced the FCC to revisit the earth station issue. In the *American Broadcasting Companies* case,[6] the use of small earth stations, approximately four to five meters in diameter, as compared with the nine-meter stations previously the norm, was found to be consistent with domestic satellite policy if certain technical criteria were met.

The policy issue confronting the FCC was how to balance the need of the CATV industry for large numbers of small, inexpensive earth stations and efficient satellite orbit and spectrum usage. Generally, the smaller the earth station, the farther apart adjacent satellites must be because small antennas do not focus their signals as much as do large antennas. The FCC permitted the use of small earth stations, finding that their use would not adversely affect the orbital spacing between satellites.

The decision was significant in two other respects. First, it continued, albeit for a short time, the attempt to specify the performance of receive-only earth stations. Second, it raised the issue of whether or not receive-only earth stations should be licensed. Within three years, the FCC would change these policies.

Rates

The evolution of the FCC's move toward deregulation is most visible through its actions on the rates of domestic satellite carriers. An understanding of deregulation, however, requires an appreciation of the traditional regulatory approach toward rates.

The Communications Act of 1934 has three sections of primary importance to the rate regulation of common carriers. Section 201*(b)* provides that all charges, practices, classifications, and regulations for common carrier communications services shall be just and reasonable. Section 202*(a)* makes it unlawful for any common carrier to make any unjust or unreasonable discrimination in its charges, practices, classifications, and regulations. And Section 203 requires every common carrier to file with the FCC all charges for services showing the classifications, practices, and regulations affecting such charges.

The language of Sections 201*(a)* and 202*(b)* is necessarily general, since it was intended to cover the broad spectrum of possible rate levels and structures that would be filed over many years. Further, it was built on the framework of cases decided under equivalent provisions in the Interstate Commerce Act. Under Sections 201*(a)* and 202*(b)*, the FCC held that the cost of service was to be the benchmark in the evaluation of the lawfulness of rates and that "like" services under like circumstances must be priced in the same manner. Rates based on value of service (i.e., market-based rates) were rarely permitted.

As already noted, satellite carriers were first authorized under a policy that would encourage them to undertake service and technical innovation and provide an impetus for efforts to minimize costs and charges to the public. This policy was followed in a series of cases involving new and often innovative rate proposals. In each instance, the FCC reiterated its satellite policy and, at most, set the rates for investigation.

During this period the satellite industry was extremely competitive. The new carriers had just launched expensive satellites and were anxious to fill them with traffic. As a result, rates were constantly moving lower and often below cost. The FCC's reaction (not surprising for a regulatory agency) was less antagonistic than if the rates had been moving higher than a "just and reasonable" level, as traditionally calculated.

This regulatory attitude was expressed in a 1975 case involving the private line rates of American Satellite Corporation (ASC).[7] Western Union protested the ASC rate decreases, but the FCC refused either to suspend the effectiveness of the rates or to set them for investigation. Instead the Commission said that ASC was a new carrier engaged solely in specialized communications and not "monopoly" services such as those provided by AT&T. In addition, the FCC said that there was no public interest justification for requiring that rates be fully compensatory at all stages of a satellite company's development. Indeed, the fact that the rates might be too low was found acceptable:

"In fostering the development of satellite, as well as terrestrial, specialized common carriers we recognize that some might not be profitable initially and that some might fail."[8]

The next several years showed a continuation of the intense competition among satellite carriers. The trend for both private line and transponder services was toward offerings for fixed periods of one year or more at lower rates than for service offered on a monthly basis. In response to these tariff offerings, the FCC investigated them when protested by other carriers. By mid-1978, the tariffs of ASC, Western Union, and RCA Americom had received this treatment. The FCC then consolidated the various cases and said that rather than dealing with them on an individual basis, it planned to institute a broad rulemaking proceeding that would consider the adoption of guidelines for competitive tariff offerings by specialized, domestic satellite, and value-added carriers.[9] This set the basis for the *Competitive Carrier* rulemaking proceedings that spanned the mid-1980s.

THE MIDDLE YEARS, 1979–1983—MORE SATELLITES AND DEVELOPING DEREGULATION

Satellite Authorizations and Orbital Assignments

By the beginning of 1979, six of the nine satellites originally authorized had been launched. During 1979 two additional satellites were launched, and a third was a launch failure.

The period 1979 through 1983 saw a rapid growth in the demand for satellite service. This growth resulted in a series of applications for new satellites, which were acted upon by the FCC in 1981.[10]

In making these grants, the FCC restated its policy that, when authorizing the initial entry of a carrier into the domestic satellite field, it would permit the launch of two satellites. The first was to be used for regular service and the second for growth, backup, and restoral capacity and preemptible traffic when it was not otherwise occupied. The FCC said that it was reluctant to authorize additional satellites without a substantial showing that present traffic and firm commitments for service would soon exhaust existing facilities. Finally, the FCC stated that it anticipated beginning a general proceeding that would investigate methods of accommodating more satellites in orbit than was then possible.[11]

The FCC started the rulemaking proceeding in 1981, asking for comments on a proposal to adopt 2° orbital separations between domestic satellites. In 1983 it decided this proceeding and, at the same time, authorized the construction of 19 additional domestic satellites and the launch of 19 new or previously constructed satellites. As a result of these and prior authorizations, the orbital assignments as of June 1985 are shown in Figure 3–2.

FIGURE 3–2
June 1985—Operational and Authorized Satellites

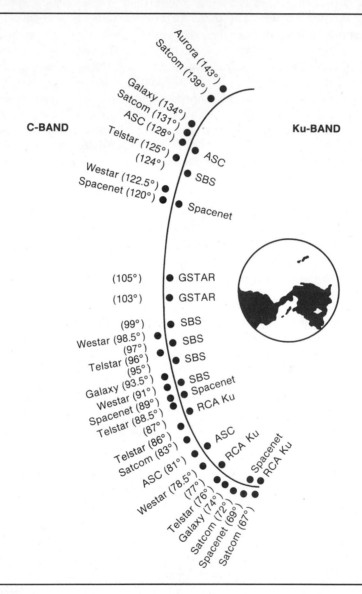

C-BAND

Ku-BAND

Aurora (143°)
Satcom (139°)
Galaxy (134°)
Satcom (131°)
ASC (128°)
Telstar (125°)
(124°)   ASC
                          SBS
Westar (122.5°)
Spacenet (120°)
                          Spacenet

(105°)   GSTAR
(103°)   GSTAR
(99°)   SBS
Westar (98.5°)   SBS
(97°)   SBS
Telstar (96°)
(95°)
Galaxy (93.5°)   SBS
Westar (91°)   Spacenet
Spacenet (89°)   RCA Ku
Telstar (88.5°)
(87°)
Telstar (86°)   ASC
Satcom (83°)   RCA Ku
ASC (81°)   Spacenet
Westar (78.5°)   RCA Ku
(77°)
Telstar (76°)
Galaxy (74°)
Satcom (72°)
Spacenet (69°)
Satcom (67°)

The main points of the orbital spacing decision were:

A spacing of 2° is the long-term orbital spacing criterion at C-band, with immediate spacings set at a combination of 3°, 2.5°, and 2°; K-band spacings would be 2°.

Two orbital locations are initially assigned to a newly authorized system. Additional locations are assigned only upon a showing that in-orbit satellites are essentially filled.

Applicants for hybrid satellites must demonstrate orbit and frequency reuse efficiency comparable with the then-current state of the art for two single-band satellites.

The upper limit on in-orbit spare capacity for any licensee would be set at the equivalent of one spare satellite used for occasional or preemptible services within the system.[12]

The majority of the satellite authorizations were in accordance with this past practice, but there were three exceptions. The FCC found that several of the applicants had committed insufficient financial resources to build and operate a domestic satellite system. These applicants were conditionally granted authorizations, provided that they showed within a specified time success in obtaining financing. This attempt at varying the policy of only granting entry to qualified entrants was to become a failure.

Simultaneously with the authorization of the 19 new satellites, the FCC established a cutoff for processing new domestic satellite applications.[13] This cutoff provided that those applications filed on or before November 7, 1983, would be processed as a group, while those filed later would have to await future FCC action.

Earth Stations

Deregulation of receive-only earth stations occurred in 1979.[14] This action spurred the CATV industry, making the installation of thousands of inexpensive earth stations possible. It also marked the continuing trend toward less regulation.

In deregulating receive-only earth stations, the FCC implemented a policy of nonmandatory licensing. Whereas in the past an operator had been required to go through the three-step licensing process previously mentioned, the new procedure permitted the installation and operation of receive-only earth stations without any filings with the FCC. Moreover, the receive-only earth station would no longer be required to meet any technical standards. The rationale for this decision was twofold: first, a receive-only antenna by definition does not transmit and therefore cannot cause interference to another user; second, if an operator did not license its earth station, no interference protection would be afforded to it. Thus, protection from interference due to existing or planned terrestrial systems would not be afforded to the unlicensed stations.

The next major decision affecting earth stations occurred in 1983 when the FCC announced its objective of achieving 2° orbital spacings. In its action the FCC amended its Rules and Regulations pertaining to the performance standards of licensed antennas. These rule changes required receive antennas to be built so as to better disregard unwanted signals from satellites adjacent to the satellite from which the signal was being received and required transmit antennas to be built to better focus the signals on the intended satellite.[15]

The rules change provided that receive-only antennas did not have to conform with the higher performance standards generally made applicable. The FCC said it was giving these earth station owners more discretion and allowing them to perform the trade-offs between the cost of a new antenna and the quality of the signal received.

With respect to transmit-receive antennas, the FCC took a two-pronged approach. July 1, 1984, was set as the date for newly installed antennas to comply with the new technical standards. Antennas installed before that date would have to be upgraded or replaced by January 1, 1987.

Small-diameter transmit earth stations were also treated. It had been suggested that antennas with diameters in the range of 1.2 meters could operate successfully with 2° orbital spacings but would not comply with the rule standard. In this event, a waiver of the standard could be requested at the same time the engineering showing was made in the earth station license application to demonstrate the compatibility with 2° spacings.

As a result of petitions for reconsideration, the FCC postponed the July 1, 1984, date for new transmit antennas to comply with the rules. When it acted on these petitions, the FCC changed the compliance date for new transmit antennas to February 15, 1985.[16] It also specified procedure under which existing antennas could be operated after January 1, 1987, without being modified. Under the procedure, operation of the nonconforming antennas either would have to conform strictly with 2° spacing interference objectives or not cause unacceptable interference with adjacent satellites.

During this period the FCC also had the opportunity to rule on several large systems of small diameter (that is, 1.2 meters) earth stations.[17] In each case the small antennas were permitted if they showed that 2° orbital spacings would not be compromised.

Rates

During this period, the regulation of common carrier rates underwent the most significant changes since the enactment of the Communications Act in 1934. The commission initiated a series of rulemaking proceedings addressing the policy and rules of rates for competitive common carrier services and then issued six Reports and Orders implementing specific rule changes.

The *First Report and Order* classified carriers as either dominant or non-dominant, depending upon their power to control price in the marketplace.[18] Carriers found to possess market power were classified as dominant, and the FCC continued to regulate them under the traditional regulatory scheme, thereby ensuring that they did not exploit their market power to the detriment of the public. Those carriers found to lack market power were deemed non-dominant, and the regulatory requirements imposed upon them were streamlined or eliminated. Under the streamlined rules, tariff changes that could be made on 14 days notice (rather than 90) were presumed lawful, and a party seeking suspension of a tariff filing had to demonstrate generally that the injury to competition from allowing the tariff proposal to become effective would be greater than the harm to the public from depriving it of the service proposed.

Using this framework, domestic satellite carriers were classified as dominant. The FCC said that space segment providers possessed market power. However, the FCC was not certain that consumers would be better off with rate regulation of domestic satellite carriers laying the framework for future changes.

The *Second Report and Order* took a more dramatic step.[19] The FCC decided to forbear from regulating terrestrial resale carriers—that is, those carriers who do not own any transmission facilities but instead obtain basic communications services from underlying carriers for purposes of resale. Under forbearance, resale carriers would not be required to file any tariffs with the FCC. Instead, their relationships with their customers could be pursuant to private contracts. Even though tariffs did not have to be filed, the resale carriers still had to comply with the requirements of the Communications Act of 1934 that rates and practices be just, reasonable, and not unjustly discriminatory.

Although these *Reports and Orders* did not directly affect satellite carriers, two significant events occurred that did affect the rates that satellite companies could charge. First, RCA Americom proposed to establish a uniform lump-sum charge of $13 million, payable in advance, for transponder service. This rate was considerably higher than RCA Americom's then-prevailing transponder rate. As support for the rate, RCA Americom argued that the rates were based on market demand.

The FCC set for inquiry the proposition that satellite carriers should be allowed to increase the price of transponders above cost to levels commensurate with demand.[20] It deferred the inquiry, however, stating that the question was already under active consideration in the competitive carrier rulemaking proceedings.

The second significant event involved the proposal for transponder sales. In late 1980 the FCC had received a letter from an existing satellite carrier stating an intention to sell ownership rights, on a noncommon carrier basis, to satellite transponders. After soliciting public comments on this and other proposals that were filed, the transponder sales were permitted as being consistent with FCC policies fostering multiple satellite entry.[21] The FCC said transponder sales would encourage additional entry, additional facility investment, and more efficient use of the orbital and frequency spectrum and would allow for technical and marketing innovation. As a result of finding that such sales were not common carriage, the rates for the sales were left entirely to the marketplace.

The *Fourth Report and Order* was adopted in 1983, and it continued the deregulatory trend.[22] For the first time, domestic satellite carriers were classified as nondominant, permitting tariffs to be filed on 14 days notice with no supporting cost data. In addition, the FCC found that allowing the rates of domestic satellite carriers and resellers to equal market-clearing prices would promote the public interest. The FCC also applied forbearance to specialized common carriers, such as MCI and domestic satellite resellers.

The *Fifth Report and Order* followed in 1984 and applied forbearance to domestic satellite carriers.[23] However, challenges to the FCC's forbearance policy were made in court by parties arguing that the Communications Act required the filing of tariffs by common carriers. Carriers also refused to withdraw their tariffs on file with the FCC even though the FCC had authorized them to do so.

The issue of whether or not a carrier subject to forbearance could refrain from withdrawing its tariff was addressed in the *Sixth Report and Order.* [24] It ordered that forborne carriers cancel their tariffs and not file any others. Carriers were given a six month transition period beginning in February 1985. The FCC action was appealed, and a stay of the detariffing order was issued by the court pending its consideration of the appeal.

DEREGULATION TODAY?

Satellite Authorizations and Orbital Assignments

In May 1985 a Notice of Proposed Rulemaking was issued in which the FCC proposed to adopt new rules specifying revised financial qualifications and transponder loading requirements for applicants seeking authorization of new satellites.

The FCC proposed that each applicant for a satellite submit (1) audited financial statements showing that it or its parent had sufficient uncommitted funds to cover the estimated investment and first year's operation cost and (2) proof that funds are firmly committed to provide all capital expenditures.

With respect to transponder loading, it proposed to continue its policy of assigning qualified entrants up to two orbital slots in each frequency band without a definitive showing of future transponder loading. For existing satellite licenses, the applicant was required to show that the additional satellites requested would be 80 percent filled within three years of launch. If no contracts for future capacity had been signed, the 80 percent showing could be made by projecting historical growth rates from the applicant's existing system. If this showing could not be made, the FCC would permit one additional satellite to be built subject to demonstration prior to launch that the in-orbit satellite system of the applicant was 80 percent filled.

As is apparent, these proposed rules continued the basic policy adopted with the first domestic statellite authorizations in 1973. The FCC proposed to license all qualified applicants until such time as there were insufficient orbital positions available to accommodate this approach. The rules are more strict than previously applied because of the orbital slot shortage facing the FCC.

As a result of the November 1983 cutoff order, the FCC had before it the applications shown in Table 3–1. The number of orbital positions available were insufficient to handle the applications, and some means of limiting the

TABLE 3–1
New Satellite Applications Filed November 7, 1983

Applicant	C-Band	Ku-Band	Hybrid
Alascom	2*		
American Satellite			2*
Columbia Communications Corp.			2
Comsat General		3	
Digital Telesat	1	2	
Equatorial Communications	2*		
Federal Express		2*	
Ford Aerospace			3*
GTE Satellite Corp.		1*	
GTE Spacenet			1
Hughes Galaxy	1	3	
Martin Marietta		2	
National Exchange	2*	2* (4 satellites)†	
Rainbow		2*	
RCA Americom			3‡
SBS		1	
Systematics General	2 (TDRSS)		
USSSI		2*	
Western Union	3§	3	*
Total	13	23	11

* Plus a ground spare.
† Four satellites would occupy two orbital slots utilizing spotbeam frequency reuse.
‡ Or alternatively three C-band and three Ku-band satellites, at RCA Americom's option.
§ Western Union has also requested authority to construct replacement satellites through year 2000.

number of applications granted was required. The proposed rules were the FCC's answer.

The orbital availability was helped slightly when the FCC voided the authorizations conditionally granted in 1983 to three companies. The reason for this action was that the companies had not shown they were financially qualified in their own right.

If the FCC does adopt the proposed rules, and the rules are insufficient to winnow the list of applicants to the number of orbital slots available, the commission may attempt to use alternative allocation devices. Devices might be a lottery, which uses random selection, or an auction, which would assign orbital positions to the highest bidders.

The remaining issue of importance to Ku-band satellite applicants and users affecting the orbit concerns the power limitation on transponders. Transponders at C-band originally had 5 watts and over the years increased to the 8.5- to 10-watt range. This power has proved to be sufficient and approximates the permissible limit.

At Ku-band, higher powers are more important because this frequency is more susceptible to signal degradation due to rain. In addition, the other attributes of Ku-band—freedom from terrestrial interference and inherently smaller receiving antennas—have made higher-powered satellites attractive.

The first Ku-band satellites were in the range of 20 watts. RCA Americom then filed for 40, and later 45, watts. This application was approved by the FCC.

The limit on Ku-band power has not yet been tested. A restriction does exist that the EIRP at Ku-band per television signal cannot exceed 53 dBW when the transponder is being used for "broadcast satellite services."[25] The FCC has not yet decided whether this limit will apply to other television users. In addition, the FCC has not ruled on transponder powers higher than 45 watts, although they are possible even within the 53 dBW limit.

Rates

The remaining issues involving deregulation start with whether or not the courts will permit the FCC to require common carriers subject to forbearance to remove their tariffs on file with the FCC. If the FCC is upheld, then future relations between carriers and customers will be by private contract.

If private contracts become the norm, the evolving issue will be the extent to which the FCC will permit itself to become involved in allegations that carriers' rates are too high or discriminatory. The current FCC position appears to be that such challenges to rates will be entertained only in the most egregious circumstances. The maintenance of this position could lead to an extremely competitive marketplace with widely varying rate levels and structures.

Another major issue facing satellite carriers involves the regulation of AT&T, which is the last dominant domestic carrier. The effect of completely deregulating AT&T is not yet fully known, but deregulation certainly has the potential for having a significant impact upon pricing.

■ CONCLUSION ■

The satellite industry has had a short history but one filled with dramatic changes. The skies have been filled with satellites under a policy of encouraging entry and competition, so that today there is more than enough capacity for every use. Earth stations have multiplied as the FCC has permitted smaller sizes and freed receive-only antennas from regulatory requirements. Technical standards for transmit antennas have become more stringent in order to reduce the spacings between satellites, but flexibility in licensing large networks of such antennas has been evidenced. Regulation of rates by the FCC during this same period has gone from a cost-based system of evaluating the lawfulness of rates to a system in which market-based rates are accepted.[26]

The challenge today for both satellite carrier and satellite service user is to balance respective needs through mutual agreement without reliance on a regulatory agency.

■ NOTES ■

1. Report and Order, 22 FCC 2d 86 (1970).

2. Memorandum Opinion and Order, 34 FCC 2d 1 (1972).

3. American Satellite Corporation, 43 FCC 2d 348 (1973); American Telephone and Telegraph Company, 42 FCC 2d 654 (1973); Communications Satellite Corp., 42 FCC 2d 677 (1973); GTE Satellite Corporation, 43 FCC 2d 1141 (1973); RCA Global Communications, 42 FCC 2d 774 (1973); Western Union Telegraph Company, 38 FCC 2d 1197 (1973).

4. RCA Global Communications, Inc., 58 FCC 2d 656 (1976).

5. Florida Cablevision, 54 FCC 2d 881 (1975).

6. American Broadcasting Companies, Inc., 62 FCC 2d 901 (1977).

7. American Satellite Corporation, 55 FCC 2d 1 (1975).

8. Ibid., p. 2.

9. RCA American Communications, Inc., 69 FCC 2d 426 (1978).

10. RCA American Communications, Inc., 84 FCC 2d 633 (1981).

11. Domestic-Fixed Satellite Service, 88 FCC 2d 318 (1981).

12. Licensing of Space Stations in the Domestic Fixed Satellite Service, 54 Rad. Reg. 2d 577 (1983).

13. Domestic Fixed-Satellite Service, 93 FCC 2d 1260 (1983).

14. Deregulation of Receive-Only Domestic Earth Stations, 74 FCC 2d 205 (1979).

15. Licensing of Space Stations (1983).

16. Reconsideration of Licensing of Space Stations in the Domestic Fixed Satellite Service, FCC 84–487 (released January 9, 1985).

17. Schlumberger Technology Corporation, Mimeo No. 4658 (released June 7, 1984); Equatorial Communications Services, Mimeo No. 2831 (released March 13, 1984).

18. First Report and Order, 85 FCC 2d 1 (1980).

19. Second Report and Order, 91 FCC 2d 59 (1982).

20. RCA American Communications, Inc., 89 FCC 2d 1070 (1982).

21. Domestic Fixed-Satellite Transponder Sales, 90 FCC 2d 1238 (1982), *aff'd. sub. nom. Wold Communications, Inc.* v. *Federal Communications Commission,* 735 F. 2d 1465 (DC Cir. 1984).

22. Fourth Report and Order

23. Fifth Report and Order

24. Sixth Report and Order

25. GTE Satellite Corporation, 90 FCC 2d 1009 (1982).

26. Regulatory changes have occurred subsequent to the completion of this chapter and will be reflected in the next edition of this book.

Author

Carl J. Cangelosi
Vice President and General Counsel
RCA American Communications, Inc.
Princeton, New Jersey

Carl J. Cangelosi is vice president and general counsel of RCA American Communications, Inc. (RCA Americom), a domestic satellite common carrier. His responsibilities include supervision of the legal and regulatory affairs of RCA Americom. Before joining RCA Americom, he held various legal positions at RCA Global Communications, Inc., in New York and Washington, D.C., Cangelosi started his legal career with the Federal Communications Commission as an attorney working on domestic and international matters. He received an A.B. degree from Georgetown and a J.D. degree from Boston College Law School.

Chapter 4

Teleports: What's It All About?

Kenneth A. Phillips

Gulf Teleport, Inc.

Outline

IS IT A REAL ESTATE OR
COMMUNICATION VENTURE?

WHO WILL USE THE SERVICES?

MEETING MARKET NEEDS

FIRST MILE/LAST MILE

CONCLUSION

Source: Reprinted from Alan D. Sugarman, Andrew D. Lipman, and Robert F. Cushman, eds., *High Tech Real Estate* (Homewood, Ill.: Dow Jones-Irwin, 1985), © Dow Jones-Irwin, 1985.

The jury may still be out on the question "What is a teleport?" The American Teleport Association organizing committee agonized over a definition of *teleport,* and the question was tabled more than once. Eventually a definition evolved which includes the following items as essential ingredients of a teleport.

Provides satellite up- *and* downlink capability (eliminates receive only).

Has some form of terrestrial link to a market area (fibre optic, microwave, etc.).

Provides more than one class of service, voice video, or data (eliminates video-only uplinks).

Provides service to multiple users on a commercial basis.

Within that broad definition there is still room for many variables, primarily in scope and level of service. In theory a teleport operator could provide two-way access to one or two satellites, provide a video channel to one major subscriber, sell a small number of audio or low-speed data circuits, and qualify as a legitimate teleport. This service could be provided to one building with relatively few users and would still fall within the scope of the proposed definition. The jury is still out in the sense that economics will eventually dictate who is the real teleport.

The broader scope of teleports is the "gateway" concept; and it appears that most developing teleports see themselves as full-service providers of voice, video, and data communications on a regional level. Therefore the teleport becomes the communications gateway to a region. The service area (region) may be one large commercial building (the National Press Building, Washington, D.C.), a metropolitan center (the New York Teleport), a geographic region (Gulf Teleport, Inc., Houston), or a statewide network (the Ohio Teleport). In each of these cases, the teleport will provide satellite access and interconnect capability to a customer base. The market potential of each geographic area must be the single most important factor in determining the level of service offered. Almost as important is the business objective of the teleport developer; specifically, is the teleport a real estate development or a communications project?

Teleports are being developed with direct real estate participation and involvement, and teleports are being developed with no real estate plans. Teleports can also be service providers to unrelated third-party real estate projects, buildings, or business parks.

The New York Teleport is definitely a part of a real estate development; and certainly teleport capability can be utilized to enhance a real estate project.

When the teleport/real estate project in New York is properly planned and marketed, there will be communication users who will wish to locate at the site of the satellite communication center.

The Bay Area Teleport has a similar real estate involvement, as do several others.

However, several potential teleports are serious communications projects without a real estate involvement. The Seattle Teleport is a project of Pacific Telecom and has no real estate plans. Gulf Teleport, Inc. has no real estate plans. The Ohio Teleport has no direct real estate interest, but one of its equity partners is a real estate developer who may potentially benefit from land development at or near the teleport sites.

Similar to the Ohio Teleport, Gulf Teleport will potentially benefit unrelated real estate developers. The map in Figure 4–1 shows the proposed microwave network through which Gulf Teleport will access its intended market area. At the time of this writing, a preliminary agreement had been reached with an unrelated third party whereby one of the teleport microwave trunk lines would terminate in a 3,000-acre industrial/commercial development. The benefit to the developer is the enhancement of the new development, and the teleport receives an expanded customer base and favorable terms for access.

IS IT A REAL ESTATE OR COMMUNICATION VENTURE?

For the real estate developer who is not experienced in communications technology, there are many traps. The technology is not simple and is rapidly changing. The marketplace will no longer tolerate unreliable service. For data communications, 99.99 percent reliability is the minimum acceptable design goal. Personnel *truly* qualified in communications technology are rare. A communications-enhanced business park could face economic ruin because its communications network proved unreliable. Great claims are made in the marketplace with new technology; many are never realized.

It is the position of this writer that the communications venture must be sound—technically and economically. If the venture is sound, a real estate development is not essential to success; however, an equally sound real estate venture can provide a mutually beneficial marriage.

The real estate developer must realize that in a service network like that planned by most teleports, a user located at the teleport site has no real technical advantage over a user remote from the site. Any advantage is limited to the marketing skills employed to enhance the real estate and is largely a matter of "image" or "cosmetics." The real key to using the teleport to full advantage will be the in-house networking capability of the real estate developer or tenants.

Without treating in depth the subject of smart buildings, it is fairly obvious that any building or commercial/industrial development that can deliver tele-

FIGURE 4-1

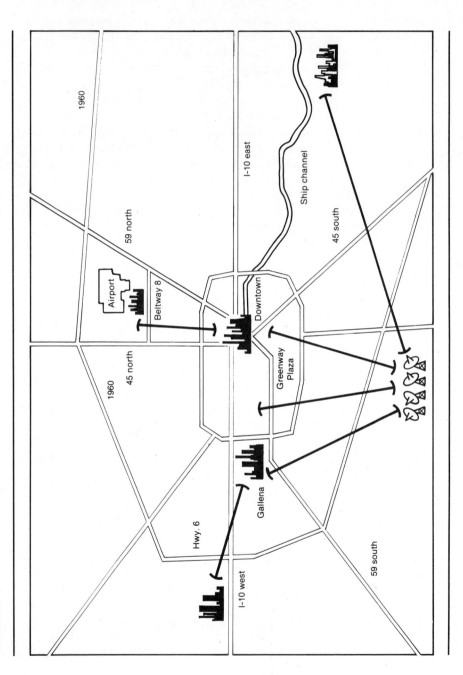

port services to its tenants can acquire a substantial advantage over competitors who have no such capability.

WHO WILL USE THE SERVICES?

For a real estate teleport, the questions, Who are the expected tenants? and What is the market? are the same. A major danger in high tech enterprises is the tendency to let technical capability be the driving force in market planning, rather than addressing what the market wants to buy. There are many technically sound endeavors that have not been able to reach profitability due to a failure to realize that the market is unwilling to pay the price requested by the entrepreneur for a technical skill or service that is dear to the entrepreneuring heart. SBS stands out as a business in trouble because the marketplace would not respond to an imagined need. Likewise, videotext endeavors are yet to reach market acceptance as expected in spite of millions of dollars spent in technical development and marketing. It appears that DBS service is headed in the same direction.

Every marketplace is unique and must be addressed with its own unique character in mind. Failure to recognize that the Houston, Gulf Coast market may differ drastically from New York City would produce a planning disaster. Therefore, the Houston marketplace will be described to illustrate the point.

Real estate developers in Houston have been slow to react to the rapid development of communications technology. Houston has only one smart building, and it has achieved only moderate notoriety. There are a few industrial companies with very sophisticated communications staffs and capability, but they are very few in comparison with the total market. One major real estate developer executive told this writer, "All we know about communications is that we always provide a telephone equipment room and a vertical chase through the building, call the phone company, and they do the rest. But they won't do that anymore." Many companies have communications managers whose historical job function was to call the local telephone company to order telephone service, either new or changes. One major national company communications manager recently acknowledged he had never heard of broadband networking. Many potential users of teleport services think 9,600 bps is high-speed data transmission. This seeming lack of sophistication is largely due to the fact that for most users, 9,600 bps or slower was the only service available in dedicated circuits from the local telephone service company.

Gulf Teleport market research revealed a need to address both markets, the very sophisticated and the unsophisticated. However, the techniques for sale of the service and for technical delivery of the service are radically different. Teleport planners must also recognize that many of the more sophisticated users may very well proceed with private networks and may not even become customers.

Another problem facing the real estate/teleport developer, especially the developer of smart buildings for teleport interconnect, is the difficulty of being all things to all people.

Management Information Systems Week reported on Tabor Center by Williams Brothers Realty Developments, Inc. of Tulsa, Oklahoma. This project apparently plans a fiber optic network in 3.1 million square feet of office space connected to an in-house computer. The company was said to be holding off a decision on computer selection until the last minute to see which system was the "most sophisticated." Certain questions become obvious: Will the most sophisticated system available translate to sellable economics? How many potential clients really need the most sophisticated system? Can the building provide a savings to its tenants, or must they pay a premium for the most sophisticated system? Can individual building owners and operators provide technical support for such systems? Each project must find the answer to these questions if it is to succeed.

MEETING MARKET NEEDS

It is therefore suggested that a properly designed teleport facility should provide communication services to the broadest possible market. The communications links used—whether fibre optic, microwave, or some other medium—should be as nearly as possible a transparent conduit for the users' signals. Whatever is sent must be delivered to the other end of the pipe (conduit) in the same form as transmitted. Gulf Teleport market research provided one overriding feature of potential customers. They said in effect, "We want to keep on doing what we are doing, do it better, and do it cheaper." Any project that cannot meet that objective may well be doomed to failure in the marketplace. Only after that immediate objective is met can the project planner expect to be a force in gradually moving the customer to more sophisticated and more expensive communications systems. The entire teleport concept has been based on the concept that by shared service, the customers could achieve greater levels of service at lower cost. Technical brilliance must serve that goal to be a success.

In many cases less sophisticated techniques are more than adequate to meet the customer's needs. *Fibre optics* is the new buzzword in communications, but in many cases coaxial cable or twisted pair technology may meet the needs of the marketplace with better economics to the end user. Trade journals and newspapers frequently report on exotic, sophisticated functions being planned for smart buildings. Again the marketplace must be examined carefully by real estate and communications planners to avoid indulging their own space-age technological fantasy.

Prospective users of teleport or communication-enhanced properties are limitless. Any company with a computer in one location communicating with a computer in another location is a potential customer, assuming that the teleport can provide superior communications capability at competitive rates. Do not be fooled, however, into believing that users will flock to a new service in a frivolous manner. Data communication is serious business to any potential customer. Unless a teleport marketer can show both quality of service and savings, the market will not respond. No matter how poor the current service, customers know what they have. No corporate communications manager will risk his/her job by recommending a new and unproven carrier, unless there is great motivation; and that motivation must be both service and cost. When those requirements are met, the market will respond.

Leaving the data world behind, voice and video users are next to be considered. Gulf Teleport planners were pleasantly surprised at the large number of dedicated leased voice circuits reported in market research. In most cases, dedicated voice circuits were in use paralleling data circuits.

Gulf Teleport anticipates that leased voice circuits will total approximately 30 percent of its total traffic. The buyers of those circuits are expected to be the same buyers as for data circuits. Video conferencing is such an insignificant part of the market that it has been totally ignored as a planned revenue base. The marketplace has yet to determine what real videoconferencing is, and many services sold as videoconferencing do not deliver true two-way video and audio. The most common form of videoconferencing is a video and audio program in one direction only. Conference attendees are congregated in meeting rooms equipped with TVRO earth stations. The conference attendees may address questions to the program distributors over 800 WATS lines.

Industry has a need and a legitimate use for true two-way videoconferencing on an occasional basis, so that both executive, management, and technical meetings can be conducted as needed. The cost savings for a large company would be significant if managers scattered across the country could confer on a real time basis without traveling to a central site. Very few companies can afford to install such a system for their private use, though that market may grow as the cost of technology continues to decrease. Teleports could make real-time videoconferencing a reality on a shared basis if compatible equipment were used in numerous teleports. The teleports would then become a vehicle for the growth of the true videoconferencing market. It is very difficult to believe that videoconferencing can be a successful industry without some shared service medium, such as teleports. Gulf Teleport, Inc., in its technical planning, expects to make such service available on a very carefully planned basis—to avoid great capital expenditure with very little return. Commitment of capital must parallel the growth of the market.

Real estate planners may find a marketable commodity by providing videoconferencing facilities in a development, connected to a teleport. Careful market and cost evaluations would be needed to ensure the economic success of such a facility. Not to be ignored is the possibility that such a facility may never be self-supporting, but its existence could enhance the real estate project to such a degree that it becomes desirable.

In order for teleports to succeed, satellite uplink capability must be combined with the terrestrial link that delivers service to the customer. There is no reason to believe that any teleport operator could build a facility providing one or more satellite uplinks and downlinks, advertise the satellite capability, and expect the market to respond. The problem is one of accessing the uplink and the downlink with good economics and good reliability. There are numerous earth stations in operation and available at the present time. Most are operating at significantly less than full capacity. The marketplace has no need for additional uplinks and downlinks. What is needed and that which will ensure the success of the teleport concept is the up/downlink tied to the terrestrial network for access to customer premises. Therefore, the terrestrial link becomes the real key, rather than satellite capability. From a marketing viewpoint, the positioning of the terrestrial link becomes the key, so that it accesses market concentrations sufficient to produce a sustaining revenue stream. Remember, teleports have an economic advantage by the very nature of the business. The teleport is not a service provided to every home or even to every business in its service area. The target market is leased line users, and the teleport planner enjoys the luxury of accessing only selected geographic areas where adequate business potential is identifiable. Therefore, with proper planning, capital is placed only where the return is assured.

The cost of a teleport is not a small item; but in terms of continuous operation, the cost of transponder rental can overshadow the cost of capital equipment. More than one uplink operator has gone broke because the revenue base would not support the cost of transponder rental.

Once a local teleport network is in place, be it on the Gulf Coast or anywhere else in the world, the part that is missing is the other end of the link. One telephone handset with no one to talk to is worthless. Therefore, a teleport operator must be able to deliver end-to-end service. The customer will not respond to and will not buy half a circuit. Therefore, growth of the teleport industry becomes important to the success of individual teleports. This is not to say that teleports will communicate only with other teleports. The teleport and its customers could communicate with another teleport and its customer base, with individual companies who have up- and downlink capability, or with other common carriers who may distribute service within a region by other means. However, in order to make the end-to-end connection, teleport operators must have working relationships with the other half of the circuit. Again, we highlight the marketing aspects. It would be impossible to sell service to a customer in Houston or New York and only deliver the service to the satellite and tell the customer to arrange the other half of the circuit. It is hoped that the American Teleport Association will prove to be a vehicle or mechanism whereby the various teleport operators can exchange information, both technical and marketing. The sales representative in Houston must be able to price and coordinate the service hookup end-to-end, whether within the local loop or around the world. Without a cooperative effort among teleport operators, the overall concept cannot succeed.

The specifics of the terrestrial link—whether it is fiber optics, microwave, cable TV, or other means—are secondary to service integrity. In many cases economics will dictate the exact medium. Gulf Teleport is planning a combination of both digital and analog microwave, allowing the needs of individual market areas to dictate the specifics. It is very possible, indeed probable, that certain locations may have both digital and analog capabilities. The objective is to provide the highest possible service capability at a reduced cost to the user.

FIRST MILE/LAST MILE

Still another issue to be resolved in teleport planning is commonly referred to as the first mile/last mile problem. As can be observed in Figures 4–1 and 4–2, which show the Gulf Teleport service network, the teleport will bring service into business areas. This is typically the case with all the teleports planned as regional gateway systems. Teleport developers often state their intention of providing service to the end users end-to-end, but what they really mean is making the service available in a specific business community. The various microwave links or fibre optic links will have head-end points or nodes scattered throughout the service area whereby individual users can make interconnects to the teleport network. How service is carried from the head-end node to individual customers is an open-ended problem requiring many solutions. There are various technologies available to solve that problem: short-range microwave, radio telemetry, broadband cable, or leased telephone circuits. Individual teleport planners must examine their individual markets and identify the most reliable and economically sound methods of individual customer access. It is very possible and even probable that some combination of all the above methods, and possibly others not even listed, can be used. As with any chain, it is only as strong as its weakest link. Therefore, teleport operators may in some cases find themselves dependent on other service providers for the first mile/last mile hookup. This dependency can certainly be increasingly important and serious to the future potential of teleports. Only if this problem can be successfully resolved so that the customer is provided the transparent pipe, end-to-end, can service integrity be ensured.

Gulf Teleport is considering the regional, or geographic, arrangement to the business community as it addresses this problem, and the problem is less severe to Gulf Teleport than it might be to other teleports because of the way the business community is concentrated in several independent and remote blocks or groups. Therefore, each node that terminates in a business

FIGURE 4-2

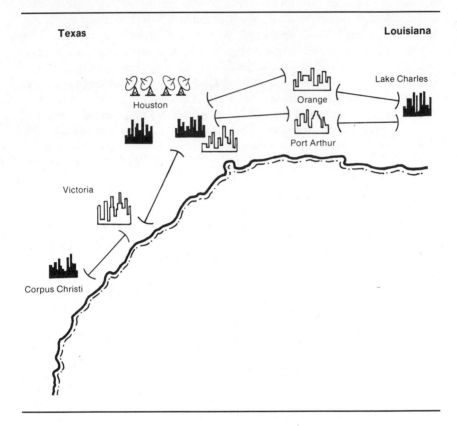

area can access a significantly large business community with little difficulty. The local telephone company (BOC) could be a help or a hindrance in solving the first mile/last mile problem. The service capabilities of the phone company could be integrated for short-range leased circuits to solve the first mile/last mile problem if the level of service integrity could be maintained through the Bell loop. The question must be resolved on a local basis. Another question is whether or not the BOC will view the teleport as a competitor and refuse to be cooperative and helpful, or view it as an opportunity to participate at the local level in the growing demand for digital circuits. Perhaps the BOC will choose the latter view and will benefit from the growth of the teleport industry and will at the same time provide a reliable and economically viable solution to first mile/last mile problems. It would be advisable for any teleport planners faced with first mile/last mile problems to attempt to involve the local BOC on a favorable basis as early as possible. For the teleport that addresses only a building or a captive real estate development, the first mile/ last mile problem is insignificant, since they could typically make direct interconnect through cable networks or fiber optic networks.

■ **CONCLUSION** ■

In summation, it should be fairly obvious that there is no pat definition of a teleport. There may be several. A teleport can be what the planners and developers want it to be within the guidelines and the needs of the marketplace that it is intended to service. The objective must be to deliver satellite communication capability to the marketplace on a cost-effective basis that will prompt a satisfactory and successful market response. One teleport cannot be a financial success without other teleports being equally successful.

Certain teleports have received a great deal of publicity through the efforts of the entrepreneurs involved. Great claims have been made concerning the importance of the teleport for local future economic development, and there is little doubt that communications capability is becoming more and more important to business planners. As new businesses come into existence and older businesses expand or relocate for various reasons, communications will be increasingly important in planning. In the past, local economic development groups have advertised such things as lifestyle, transportation capabilities, labor resources, shipping facilities, airports, harbors, and even recreational facilities. Communication capability will become a part of the evaluation criteria for business planning. It would be overly optimistic to believe that teleport capability will be the overriding factor to a business planner, but certainly the existence of a successful teleport is a valid tool for local and regional development proponents. Local Chambers of Commerce, if astute, should move quickly to embrace the teleport as an economic development tool. Once again, marketing becomes the key. Only through a balanced marketing program can the teleport concept be a valid tool for regional developers. It is also reasonable to assume that if a company were evaluating potential sites for a new factory and all other things were equal, the teleport could easily be the deciding factor. Certainly the city with teleport capabilities would have a decided advantage over the city with none. This is not to say that the teleport must be in every case a full regional teleport. Teleports can take many forms, and the form of each teleport depends on the needs of its unique marketplace.

Author

Kenneth A. Phillips
Chairman of the Board and CEO
Gulf Teleport, Inc.
Houston, Texas

Kenneth Phillips is chairman of the board and chief executive officer of Gulf Teleport, Inc., a company he helped organize in 1983. Additionally, he is president of Auta-Tronics, Inc., a specialized communications and electronic systems contractor and integrator in Houston— a position he has held since 1978. Prior to that time, he spent 12 years in the energy industry as a purchasing and materials manager and as a financial planning analyst.

Chapter 5

Teleport: The New York City Metropolitan Area Intelligent Network

Robert Annunziata

Teleport Communications

Dr. Stanley M. Welland

Merrill Lynch & Co.

Outline

Source: Reprinted from Alan D. Sugarman, Andrew D. Lipman, and Robert F. Cushman, eds., *High Tech Real Estate* (Homewood, Ill.: Dow Jones-Irwin, 1985) © Dow Jones-Irwin, 1985.

THE CONCEPT

The Staten Island Teleport is a prototype office park for the 21st century with the integration of enhanced office facilities, communications access to satellites and terrestrial networks, and a regional communication distribution system for flexible and real-time communication service requirements.

The information explosion that started in the 1970s has accelerated significantly and is changing the way we live and conduct business. Statistics have shown that 46 percent of the nation's work force is employed in the information industries, and 26 percent in service functions. In the New York metropolitan area, 55 percent of all business is generated from information industries. By 1990 an estimated $100 billion will be expended annually for business communication services. Our information economy is driven by this communications explosion. Over the next 10 years, there will be a substantial increase in the number and types of communications satellites, a major increase in the optical fiber network distribution throughout the world, and the evolution of more and more sophisticated intelligent premise switches/PBXs, all of which will serve our information-dependent societies.

Integration of this communication technology explosion with information processing and the need for enhanced real estate is the basis for the Staten Island Teleport. The Teleport addresses the business needs of all corporations for:

Communications facility access within the New York/New Jersey region for satellite and terrestrial services.

Rate stability and predictability for communication services and real estate for controlling long-term expenses.

Enhanced office facilities for controlling building energy consumption; intraoffice communications and gateway access to satellites and a regional fiber network.

Secured building office space.

Fail-safe power for protection of computers and information systems from loss of data and sabotage.

Provision of an alternate telecommunications infrastructure to meet the telecommunication needs of business in the region. The Teleport is the only digital, fiber optic telecommunication network in the New York/New Jersey area, making it the innovative option for businesses.

THE TELEPORT CONCEPT

As we approach the new century, it is becoming more apparent that the concept of a "port" with all its implications—exchange of goods, commercial transactions, and so on—must be expanded to include communications access as the critical ingredient for continued viability.

The Port Authority of New York and New Jersey, with its broad mandate to ensure the vitality of trade and commerce in its port district, has an important stake in expanding the port concept to ensure that telecommunications can find a safe harbor in this region as did the great ocean-going vessels at the dawn of this century, or as aviation does today. It was to this end that the Teleport concept was developed by the Port Authority. The idea was developed after reviewing the business needs in the New York metropolitan area and because of the particular concern about business leaving the region.

COMSAT, Arthur D. Little, and other consultants were retained to review and define this project. After much analysis, it was determined that an enhanced suburban office communications facility with transmission capabilities to the business district of New York City and the surrounding area would be viable.

Because New York City is one of the most microwave-congested urban areas in the country, any large installation of C-band earth stations and even limited use of additional microwave facilities for distribution would be very difficult. A regional site search was conducted reviewing 29 locations for satellite access to domestic and international satellite arcs. Staten Island was chosen as the best location with the least radio frequency interference. It was decided to use fiber optics for distribution from the Staten island site to and throughout the region. Fiber optics was chosen because it would avoid the microwave congestion within New York City and because of its potential capacity, which is limited only by technology advances. It is also the most secure method of transmission.

Thus development of the Teleport, to be located on 350 acres on Staten Island, was announced. The principal partners in this endeavor are the Port Authority of New York and New Jersey, Teleport Communications, and the City of New York. The Port Authority is responsible for the Teleport-site land development, including infrastructure and enhanced real estate. Teleport Communications, a joint venture between Merrill Lynch Telecommunications, Inc., and Western Union Communications Services, is responsible for the communication management of the satellite earth station access facilities, enhanced office park communications services, and fiber optic distribution throughout the region. The City of New York owns the land.

Since this conception, the Teleport fiber optic telecommunications network has evolved to meet additional communications demands of the NY/NJ region. Divestiture of the Bell System has created a major need for regional telecommunication high capacity services at quick installation intervals. This business need has reinforced the requirement for an alternate telecommunications infrastructure to serve the region. The Teleport fills this requirement.

ENHANCED REAL ESTATE "MASTER PLAN"

The Teleport's master plan (Figure 5–1) illustrates the basic physical design. This plan carefully balances the suburban office park setting with a natural wetland preserve. The development has been divided into phases: Phase I is for the development of 100 acres, and Phase II is for the development of an additional 100 acres. The 150 remaining acres will remain as an undeveloped wetland preserve. The Teleport office park will be composed of office buildings of three to four stories (80,000 to 100,000 square feet); the buildings will be clustered throughout the site and have ample parking. The entire location is fenced for security with one main entrance. The communication center, or Telecenter building, accommodates the communications switching, monitoring, and control of the earth stations within the "earth station shield" area and the fiber optic network.

Enhanced office buildings are typically used for back-office and computer

FIGURE 5–1
Teleport's Master Plan

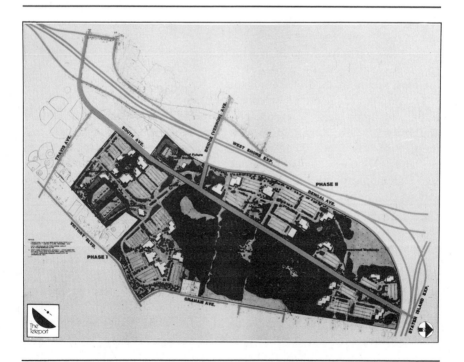

services. Computer-controlled environments with remote sensing monitors and raised floors are typical.

A unique fail-safe power system is being employed at the Teleport. The primary power is from the Power Authority of the State of New York (PASNY) on the northeast power grid, but a second contingency system serves Staten Island. This second system is an extension of the mid-Atlantic power grid from New Jersey and is provided into the Staten Island system. This system provides the fail-safe power many companies are seeking.

Security is provided by the Port Authority police, who have state police powers. There will be 24-hour coverage, seven days a week, with trained personnel for fire and emergency medical services.

The population surrounding the Teleport is important to ensure a proper labor force available for business at the site. Staten Island itself has approximately 150,000 residents with the skills, education, and training to support back-office and high tech operations envisioned at the Teleport. Within one hour of mass transportation, the same labor profile is increased to 750,000; within a half-hour automobile drive, it is expanded to 1.5 million workers.

Access to the Teleport was another critical area of concern. The site provides an excellent location. It is surrounded by an excellent highway system. Newark airport is only 12 minutes away from the site. Further, it is an easy 30-minute drive to Manhattan.

To meet the needs of the region, various land and real estate options are being made available to potential corporate tenants. Private developers are being encouraged to lease land and rent office space. Private corporations are also being offered the opportunity to lease land, and the Port Authority will finance and/or build to suit. The Port Authority makes available long-term lease arrangements and depreciation allowances; it secures local community and city approvals and arranges for city tax abatements.

TELECOMMUNICATION SERVICES

A teleport provides three essential telecommunications services: multisatellite access, a regional distribution network, and enhanced office park tenant services. From a telecommunications point of view, a teleport is a network that clusters multiple satellite access facilities away from the congested microwave airwaves of central cities. It also supplements the present earth station access facilities, which are insufficient to meet the growing needs for satellite communications. Finally, a teleport provides a tenant office park with enhanced telecommunications-enriched office facilities and ease of access to all types of communication facilities.

Teleports are a necessary addition to the communications infrastructure because forecasts predict a growing number of satellites and transponders. Forecasts indicate that C-band (4 to 6 Gigahertz) transponders will rise from an estimate of 345 in 1984 to 960 by 1990. The emerging Ku-band (12 to

14 GHz) transponder growth is exemplified with a use of 84 transponders in 1984, growing to 659 by 1990.[1]

Satellite Accessibility

With the ever-increasing number of transponders available for communication, the underlying issue in the New York region is access to these transponders. The need is not just the access to a limited number of transponders but the access to all satellites and transponders. Staten Island was selected because it is capable of accessing the full U.S. domestic satellite arc as well as the Atlantic international arc. Not only does this site provide appropriate access, but there were other important criteria used in its selection.

Proximity

Proximity to the metropolitan business district was one criteria. The Teleport is 11 airline miles away from Manhattan. This provides significant terrestrial communication cost savings for businesses and residents of the New York City region. There are approximately four common carriers that utilize C-band satellites for their communications offering in the New York region. Their satellite earth station facilities to access the domestic arc are more than 50 miles from New York City. This greater distance costs the user or common carrier an estimated 300 percent more than would be the cost of Teleport services.

An even more dramatic impact of proximity and satellite access is with international satellite services. Today, all satellite international access from the New York region utilizes earth stations located at Etam, West Virginia, or Andover, Maine. Both of these locations are more than 300 miles from New York City.

Ease of Access

Ease of access to various satellites, domestic or international, is another key element of the Teleport. Common carriers in the New York region operate their earth station facilities each utilizing the earth station to access only their own satellites. This restriction of satellite earth station access to only the common carrier's satellite places a burden on the corporate user. A user must incur charges for more than 50 miles of terrestrial lines to access a common carrier earth station and satellite in one part of the country and then may need to incur the same charges to access another common carrier facility located in a different part of the country. A Teleport user will have access to many common carrier satellites. This will reduce the need to access multiple terrestrial facilities and will provide significant savings to the user— be it a private corporation or a common carrier. In addition, this same terrestrial link will have access to international earth stations.

Satellite Services

The Teleport offers unique satellite earth station services. These services are flexibly designed to meet the needs of business. They range from protected environments for earth stations to radio frequency (RF) port-sharing services, which provide voice, data, facsimile, and video communication services. Initially, 11 acres are being used to house 17 earth stations, of maximum size of 13 meters, with room for expansion to an estimated 36 earth stations. The initial earth station is protected from stray radio frequencies with a shield wall, 50 feet high, with steel siding on the exterior and concrete on the interior.

The construction, placement of earth station, and maintenance of this area is controlled by Teleport Communications. A one-of-a-kind computer program has been designed to ensure proper placement and operation of all earth stations within the shield wall. This program calculates radio frequency and structure blockage.

Figure 5–2 depicts the Teleport shield wall with the integration of the Telecenter building and strategically located equipment shelters along the perimeter of the shield. The shield is also equipped to distribute communications facilities throughout the infield and along the wall to the equipment

FIGURE 5–2

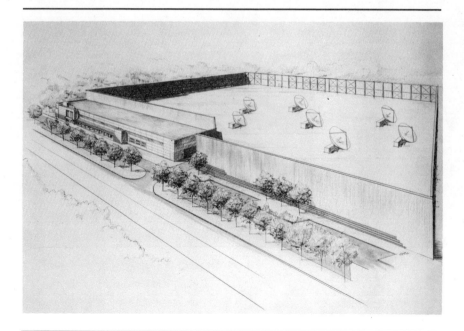

shelter. It is also designed to accommodate all environmental considerations required for installation and operation of earth stations and communication facilities. In addition, each equipment shelter will have access to the Teleport fail-safe power system and Telecenter uninterruptable power supply and generator backup.

To meet the various needs of the region and industry, the Teleport will accommodate shared or dedicated earth stations. Options are available to permit a private corporation or common carrier to install and operate its own earth station facilities. Customers can also choose to share earth stations available from Teleport Communications.

A unique offering and significant satellite communications reason for teleports is the port-sharing concept. This concept encourages earth station operators to contribute two RF ports from their earth stations to the Teleport. These ports increase the access to multiple satellites utilized by all users at the Teleport and throughout the region. (See Figure 5–3.)

OPTICAL FIBER NETWORK

Teleports involve more than the transmission and receipt of satellite communications from satellite earth stations. Paramount to a teleport is the re-

FIGURE 5–3

gional distribution network of high-capacity communication facilities. This system provides satellite users end-to-end service at various transmission speeds.

The Teleport uses fiber optical cable throughout the region. The system enables satellite services to be distributed from the Teleport to the business districts within the region. Initial installation costs are more expensive than those for microwave, but the many superior attributes of optical fiber far outweigh the initial costs. First, the microwave congestion issue is resolved. More importantly, technological improvement of optic fiber technology is proceeding at a very rapid pace. Originally a single pair of hair-thin fibers could transmit 672 simultaneous voice conversations. Further, technology has improved the quality of the fiber material and electronic terminal capacity.

Today, the transmission capacity of those hair-thin fibers is 560 Mbps, which is an equivalent of more than 8,000 voice conversations, by a conservative estimate. Improvements and advancements are expected to provide for even greater capacities to 1.2 Gbps capacity. The Teleport regional network spans the New York and New Jersey region with interconnection to the Teleport. The network provides for full redundancy of electronics and standby optical fiber pairs to ensure continued service to its users. Full physical diverse routes are employed from Manhattan to the Teleport site. An optical fiber ring network in Manhattan is designed to ensure that the system does not fail to function during potential cable outages between the strategically located optical fiber network operation centers. The initial network interconnects the Teleport with Manhattan, Brooklyn, and Queens of New York and with Newark, Jersey City, North Brunswick, and Princeton of New Jersey.

To meet the demand of users, the Teleport constructs optical fiber links from the fiber network operation centers to a user location. This ensures user service from customer premises to the satellite earth station at the Teleport and to other locations within the region. (See Figure 5–4.)

Teleport optical fiber services are offered at basic DS–1 (1.544 Mbps) and DS–3 (45 Mbps) transmission speeds. The services are offered to users anywhere in the region to the Teleport for domestic and international earth station access; from user premise to user premise; and from user premise to interstate and/or international common carrier's point of presence in the region. The optical fiber network flexibility provides the exact options that today's common carriers and businesses require to control their business communications and provides the diversity essential to ensure reliable communication services.

Today's communications environment is in disarray. With the many interstate and international common carrier services to choose from, businesses require high capacity and direct access to these companies. Direct access from the business office or real estate development to the common carrier permits cost savings and flexibility not only to the corporate user or developer utilizing communications, but also to the common carrier. Common carriers can streamline network designs and thus prefer direct access to their clients

FIGURE 5–4

via the present dominant local telephone company or an alternate supplier of local facilities. (See Figure 5–5.)

Direct access also provides a type of client control for the interstate/international common carrier to its clients by virtue of the large capacity service commitment. Direct access along with today's technology utilizing plug-in electronics (1) reduces installation intervals, (2) provides communication capacities required for various services ranging from low-speed data, voice, to video conferencing, and (3) lowers costs because of streamlined installation and network design.

The Optical Fiber Network provides the intelligent region telecommunications infrastructure to meet the business needs of the region and relieve the bottleneck of local facilities to the end users from interstate and international common carrier operating offices. It is the innovative option all businesses require for fast, diverse telecommunication services in this region.

TELEPORT PARK COMMUNICATION SERVICES

A significant response to industry needs is the teleport-enhanced suburban office park. Shared multi-tenant telecommunication services will be offered

FIGURE 5–5

to all tenants of the Teleport. Teleport tenants can avail themselves of the latest technologies now and in the future to supplement their office facilities. The Telecenter building functions as the control and monitoring center for all earth station facilities, teleport optical fiber systems, and the premises to house the intelligent systems for park services. (See Figure 5–6.)

As many people know, the shared multi-tenant system concept benefits all users by the sharing of capital costs and by achieving high-price performance. Utilizing this system will save users office floor space and the costs of preparing this space for electric, heating, ventilation, and air conditioning use by a communication system. The Teleport office park will be prewired from the Telecenter building to each park building and within the building. The prewiring plan allows for the most flexibility and is designed to provide each station user access to many voice, data, and video services. A combination of optical fiber and copper cable distribution will be utilized. Of the many park services offered, three basic systems can be described as offerings for the multi-tenant: voice management, data management, and building services management.

Voice management systems will provide basic voice communication switching service. The system will allow intelligent terminals to be connected to

FIGURE 5–6

this system provided by the Teleport or the user. Standard enhanced features such as conference calling and detailed accounting for recording of station user calls will be provided. Of significance to the voice management system will be the economic call-routing program. This program will pool all multi-tenant calling requirements and provide for the most economical medium for call completion over a variety of interstate and international carriers. The combining of the many tenant calling needs will clearly save all tenants communication costs over subscribing directly for communications from interstate/international carriers directly.

Data management systems at the Teleport will permit a tenant to again save communications costs with utilization of modem pooling and the sharing of access lines and facilities to sophisticated data networks for switched 56 Kbps and 1.544 Mbps service. With the utilization of adjunct processors, terminal emulation will be performed to allow various communications terminals and word processors to interact with other devices on the park premises or even in other cities.

Clearly the building management system will be of significance to all tenants at the Teleport. This system will utilize a protected processor for such services as central message center–electronic mailbox, electronic directory, and security alarm services. Security services will employ various devices, such as employee ID readers and sensor technology for monitoring of security entrances and doors.

Security monitoring or alarms may be transmitted to internal local building security desks or to the Port Authority police monitoring location for immediate attention of the situation. Another large application of sensor-based technology will be the centralized energy control system. This centrally protected system will save tenants from separate system installations and provide for reductions in energy consumption by controlling and monitoring lighting, fans, motors, and other electrical consumption devices throughout their premises.

Each system is designed for ease of use by individual tenants and is basically partitioned in the main system to simulate a separate control system for each user. The shared systems for tenants at the Teleport optimizes economies of scale and complements the enhanced real estate requirements of today's business community.

TELEPORT—THE INTELLIGENT REGION/GATEWAY TO THE WORLD

As Teleport puts in place a telecommunications infrastructure for the region, new regional and international telecommunication needs and pressures begin to materialize. A range of additional enhanced telecommunications and information services will be developed to utilize the telecommunications network and be housed at the Teleport office park.

As the telecommunications infrastructure becomes the core of the region's economic strength, such entities as airports, seaports, world trade centers, and business districts, will require more and more of a central infrastructure for manipulation and transmission of its information within the region and to the world. The telecommunication infrastructure will be the core of the region's prosperity and success of businesses.

Teleports will provide the access to satellites, microwave, coaxial cable, fiber optic networks, and undersea international cable systems for this service. As such, they will become the true gateway for communications to and from the region.

A teleport will take on the responsibility to meet both the national and international transmission standard to ensure service throughout the world. In addition to meeting transmission standards, world businesses are asking for transaction standards between teleports for improved efficiency between countries.

Various protocols, such as Bisync, SDLC, HDLC, ADCCP and X.25, are utilized by different businesses for their telecommunication and information processing. The characteristics of these protocols are different. Some are better suited for satellite transmission, and others are for terrestrial or

FIGURE 5–7
The Intelligent Region

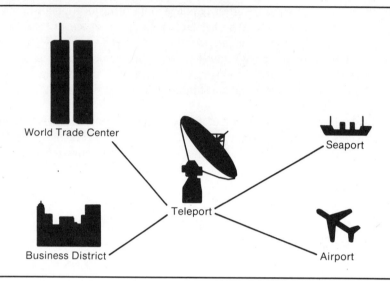

World Trade Center

Teleport

Business District

Seaport

Airport

undersea cable transmission. The Teleport will have to provide the manipulation or translation of these protocols to connect its national protocol or protocols with the proper protocol to interface with other teleports around the world. This conversion service will clearly expedite and provide ease of communications, and promote business, economic development, and regional prosperity throughout the world.

The integration of high tech communication and enhanced real estate is the Teleport's response to the New York region's businesses and economic developmental needs.

Accessibility

The Teleport site provides easy access for businesses within the region and from the regional airports. There is unparalleled access to an experienced work force on Staten Island and throughout the region. Teleport provides access to domestic and international satellites, and its distribution systems span the region and the globe by connecting users to interstate and international common carriers.

Flexibility

To meet the many varied financial and operational needs of the region, the Teleport offers a variety of service options. These services are customized building designs, various communications services and—with use of high-capacity transmission services—the quick provisioning of telecommunications within the region.

Reliability

The Teleport provides a secure environment with Port Authority police protection and a communication radio frequency-free shield. The unique fail-safe power system combining the North Atlantic and Mid-Atlantic power systems is the only one of its kind. The optical fiber network with redundant systems and physical diversity provides the insurance required for continuing communications systems.

Economy

Port Authority options of financing, tax advantages (and potential tax abatements), and attractive real estate prices provide for business use of the first New York City suburban office park. Telecommunications are competitively priced for park services and regional distribution by offering rate stability and the predictability required by all businesses. The basic philosophy of systems and satellite earth station port sharing ensures cost savings for all.

Dependability

Besides the physical and protected systems employed, the Teleport was developed in response to regional needs by reliable, qualified organizations. The project has great staying power and the commitment for success from the Port Authority of New York and New Jersey, Merrill Lynch, Western Union, and the City of New York.

An analogy can be drawn between a teleport and an airport. The essence of a teleport is an interstate/international access facility that services the telecommunications industry and other communication users in a manner very similar to the way an airport services the airline industry. In both cases, there is a sharing of common facilities at a specially developed site where the shared costs would be lower than if each user created its own site. Further, the presence of many airlines at an airport facilitates connecting flights for its passengers in the same way that the presence of many satellite earth stations at a teleport facilitates interconnections of telecommunication services. Finally, airport users share common highways into a metropolitan business district and region, and Teleport users share a high-speed optical fiber network into Manhattan and throughout the metropolitan region.

■   **NOTE**   ■

1. "Product Focus: Satellite Communications," *Information Week,* May 20, 1985, p. 44.

Authors

Robert Annunziata
President and Chief Operating Officer
Teleport Communications, Inc.
New York, New York

Robert Annunziata is president and chief operating officer of Teleport Communications, a joint venture between Merrill Lynch Telecommunications, Inc., and Western Union Communications Systems, Inc., and president and chief operating officer of Merrill Lynch Telecommunications, Inc.

"The Teleport" is a satellite communications center and office park currently being constructed on Staten Island, New York, by Teleport Communications and the Port Authority of New York and New Jersey. When completed, the Teleport will support 17 earth stations offering direct access to all domestic and some international communications satellites for transmission of high speed digital data, video, voice, and facsimile communications.

Annunziata's overall background encompasses 17 years in the communications industry with American Telephone & Telegraph Long Lines. He is experienced in operations management of switched services, voice, data, international, as well as TV and radio program services. His strong sales and marketing experience is marked by his major involvement with National Account Management in the media and financial services industry and management training development for National Account Markets.

Stanley M. Welland
Vice President and Group Manager, Technology Planning
Merrill Lynch & Co.
New York, New York

Stanley Welland is accountable for the establishment and direction of a technology planning group that provides technical planning, guidance, and assistance to corporate systems and management to maintain consistency of planning and implementation with the overall objectives of Merrill Lynch. The group constantly reviews and assesses the industry developments for potential impact on the company's information processing systems. Welland works with Merrill's venture capital organization and participates in the identification and recommendation of investments in technology that will not only yield capital growth but will also provide the firm with technological leadership in providing financial services. This covers hardware, software, and services—including satellite, cable, and related technologies. Welland joined Merrill Lynch in March 1981, when he initiated Merrill's development of The Teleport project.

For 11 years Welland held assignments in the following Exxon organizations: Exxon Office Systems Company; the Mathematics, Computers, and Systems Department; the Office of Management Information Coordination; the Headquarter's Administrative Services Department; and Exxon Enterprises (Information Systems Division). His work experience encompassed the management of a broad range of projects involving the solution of technical and business problems encountered in many of the functional areas of the Exxon Corporation and its worldwide affiliates. He was the Qyx product manager assigned to Exxon Office Systems Company and a senior adviser for the Exxon Enterprises Information Systems Division, where he was responsible for developing, evaluating, and recommending strategic and tactical plans for the growth of Exxon Enterprises' business activities in the communications systems area.

He received a Doctorate in Engineering Science from Newark College of Engineering in 1970 and an M.B.A. in finance and business systems at Rutgers Graduate School of Business Administration, 1975.

Chapter 6

Teleports in Medium-Size Markets

E. A. Eagan

Central Florida Teleport

Outline

THE DECISION TO BUILD A MEDIUM-MARKET TELEPORT

Medium-market teleports are oddities in the current teleport universe. This chapter reviews the specific problems and opportunities and the necessary preexisting conditions that will allow a medium-market teleport to survive and prosper in its chosen marketplace. The word is *chosen,* for to be successful, developers of such teleports must choose locations with great care and caution.

Many people believe that a teleport in a medium-size market is an unwise investment because a medium market doesn't have the preexisting demand for large volumes of voice, data, and video services found in a major market. However, Central Florida Teleport, one of the first medium-market teleports, was built in 1984 without the presence of such preexisting demands.

Results of a market study performed in 1983 indicated that a video uplink could be constructed and operated successfully under the following conditions: (1) An initial client could be obtained for a minimum of 200 hours of annual uplink service; (2) debt could be kept at a minimum; (3) a two-year operating budget could be established, and (4) other less difficult criteria could be met. After meeting these criteria, the Central Florida Teleport could commence operations and build up to a level of capacity designed to suit the future growth projected in the marketing study. When the initial client telephoned in July of 1984, the prior planning was complete. The initial service for that client was turned on in just less than 60 days from the date the client signed his contract.

The marketing study showed that the largest recurring costs of operating a teleport were engineering and marketing. Therefore, an additional part of the specific plan for the teleport was to offer marketing, management, and engineering services to other medium-market teleports. The complexities of construction and daily operation make shared management a cost-effective trend. Also, the high cost of equipment maintenance, marketing, and general engineering services can create a built-in failure for medium-market teleports that operate alone, since these teleports have limited revenue bases. Shared management recognizes the potential to increase profit by cutting overhead. Top-quality engineering, marketing, and office support staff can easily cost $200,000 per year. This places a substantial premium on results in a situation in which revenue is limited by market size. However, by sharing personnel, office, accounting, and general administrative expenses, several teleports can reduce overhead expense by 75 percent.

PREREQUISITES FOR A WORKABLE
TELEPORT LOCATION

The specifics of Central Florida Teleport's position in the teleport universe make it a potential model for any medium-market teleport. It had its share of problems during early development. These ranged from cost overruns of a 9-meter antenna installation to high staff costs due to higher-than-projected hours of usage.

Below is a list of requirements for ensuring success of a medium-market teleport.

Site

The site should be central to the region served and able to be frequency cleared for the entire orbital arch without expensive shielding. The terms *frequency clearance* and *frequency coordination* are covered in greater detail in the chapter on teleport engineering. Briefly stated, frequency clearance is a check of the area surrounding the proposed teleport site in search of other users of the same microwave frequencies the teleport will use. The site must not interfere with any existing microwave transmissions. In many areas of the United States, it is not easy to find a site where microwave is not already present to some degree. Erection of artificial barriers to both existing microwave ingress and egress can be a major cost factor. In the New York Teleport, creating a barrier involved erection of a 60-foot wall and was reported to have cost $4 million.

Land Availability

Central Florida Teleport is located on 300 acres that was already slated to be an office and industrial park. It is important to have enough land to expand to at least 10 times the startup size as raw land is the least costly item in teleport development. Sale of this asset, if the venture fails, can help to soften the blow of failure for investors.

Growth

The Central Florida Teleport is located in one of the top 10 population growth areas in the United States and is the second fastest growing market in Florida. Growth of this magnitude can create many opportunities for the teleport operator. Telephone, voice, and data services are needed by the new businesses locating in the market.

Development of Market

TV production companies, TV stations, universities, other schools, equipment distribution companies, direct marketing companies, and hospitals all represent important contacts to make and keep in order to be successful in developing the teleport.

Transportation: Air, Rail, Road, and Telephone Systems

Air, rail, and road transportation combine to determine the acceptability in a market for new businesses. A good set of services greatly increase the chances of attracting a new company to any given area. The Central Florida Teleport market was served by each of these transportation methods.

Telephone service depends more on who provides the service than availability. The presence of a non-Bell operating company in the market can be a key success factor. Deregulation of the telcos has caused non-Bell companies many problems in providing advanced communications services to their customers. A good relationship with the non-Bell company can be a substantial benefit to the medium-market teleport. The teleport can provide the non-Bell company many services.

Knowledge

Ten years in the communications industry and a history of acting as a promoter for many other satellite communications projects gave the author a unique window on the developing teleport industry. This knowledge provided many industry contacts that other teleport developers may not have.

Planning

The teleport required two years of preplanning and the staying power to wait two years for profitability. It is necessary to have a plan for survival and have a way to grow. How will you stay alive in the first two years? What expansion moves will you make, and when will you make them? Write out your plan and follow it. This advice is from the most basic of textbooks in starting a new business but is so vital to success.

Money

A medium-market teleport should depend upon a termination liability in its new contracts to ensure that a new capacity is profitable. The basic system installed must be nearly devoid of debt. This helps to keep operating costs in line with worst-case revenue projections.

Flexibility

The concept of flexibility is extremely important for the medium-market teleport. Implementing this concept requires plans, budgets, and delivery times for each type of service. However, one should only implement the parts that will make a profit. For example, suppose you operate a video facility and have a request for data service. You must know where to get the information on hardware cost, technical standards, and hardware configurations. To find this information, attend multi-tenant real estate and satellite industry trade shows.

Response Time

Keeping up with the delivery times of various component parts is a job that takes several hours per month. Details of the delivery time and pricing of all major components must be kept up to date and readily at hand. This information allows the manager to answer inquiries quickly and with accurate and detailed information.

Life Quality

When a potential owner begins deliberations as to whether or not a specific medium market will support a teleport, it is wise to consider carefully the existing life quality of the market. The existence of teleport services in a location with high quality of life can do much to ensure the teleport success. Companies seeking life quality for their employees will look to such a location. Local chamber of commerce officials will support the venture when they understand that it will tend to bring technology-oriented companies to the region.

Additional Management Techniques

In addition to the items listed above, Central Florida Teleport has used other management techniques and ideas, which are shown below. Although none is new, when combined with good location, they go the extra mile to ensure success.

Use "carrots" in manager compensation. If the teleport wins big, the manager wins big. This is also called incentive compensation or knowing "what makes Johnny or Sally run."

Each additional teleport capacity (i.e., voice, data, and video) can only be added after receipt of a contract containing a termination liability payment. The payment covers the cost of this new capacity to the owner if the contract is terminated before it expires.

The initial teleport staff should be added at minimum levels and minimum wages that can be supported by the sales level of that staff. Add only enough staff to do the jobs at hand.

A key management principle should be to survive and prosper over some longer time and not to get rich on the first client. It should be understood that each client must be a separate profit center if the teleport is to be successful. This concept of "phased profits" means that each service level must support its proportion of capacity and expenditure.

Summary

The medium-market teleport is a different breed of cat than the major-market teleport. Medium-market teleports are clearly desired and necessary to complete the 1990s communication system. This system will include digital microwave, vast amounts of fiber, and teleports.

The key to success is a highly flexible planning and management philosophy when implementing voice, data, video, and interconnect services. This philosophy must be coupled with an acute awareness of both the local and national markets. The one constraint is that the customer must be willing to sign a contract for the services desired that will ensure the necessary equipment can be paid for over its expected useful life.

KEYS TO PROJECT PLANNING

Aggressive, Educational Marketing

All potential users in the market area must be educated regarding the costs and benefits of using the teleport. They must be able to compare costs of using telcos and teleports. It is likely that the cost of satellite use will continue to decrease as the cost of telco circuits continues to increase. Maintain contact with the prospect so that when the two costs cross, the potential user will decide to be a client of your teleport.

The Role of Flexibility

Flexibility in its broadest sense ensures that the teleport can respond as the market grows. Keeping up with changes in technology is a big job and requires lots of reading and a constant updating of equipment files. Not enough can be said for the medium-market teleport manager who can delete or add a service quickly.

Minimizing Short-Term Debt

Carrying minimum debt will reduce short-term money pressure. A ratio of 80 percent equity to 20 percent debt is a good target. A cash flow analysis of a potential teleport market will indicate more accurately what amount of debt is affordable.

Phased Profitability

Phased profitability is a method of operation that assumes that each layer of service added to the teleport will be a profit center unto itself. This takes very careful attention when preparing the pricing for a new client. Consideration must be paid to equipment usage, staffing, and overall costs.

Interconnects within the Local Region

One must plan for future fiber and microwave routes as well as a CATV interconnect and know about the availability of all existing communications towers in the area. After ascertaining the options, determine what the costs are for using any of these interconnects. With these prices, a marketing staff can contact potential users and discuss the application of these "planned" interconnects.

Planned Office Space

Central Florida Teleport initially constructed a telecenter building to house the uplink control electronics (500 square feet). Office space was rented for marketing and operations (900 square feet). These offices are located in an industrial park about 300 yards from the telecenter. Two buildings were planned, an expansion of the telecenter to 2,500 square feet and an office building of 25,000 square feet. The office building was designed to be constructed in four stages—the first totaling 10,000 square feet and the next three stages adding 5,000 square feet each. A preconstruction leasing package was prepared and marketing will begin in the fall of 1986.

OTHER ISSUES

The viability of teleports and especially of medium-market teleports has already been covered in detail, but the question of fiber optics and its impact on the teleport industry remains to be considered. There is no agreement about what impact fiber will have on the teleport industry, but I believe that fiber's impact on teleport development will be felt most strongly in medium-size cities. The top 50 markets will all have "teleports," as in New York City. The next 50 cities will be served by fiber. The necessary fiber construction to make this theory a reality is already well under way. The Southern New England Telco fiber network in Florida is a good example. It stretches from Miami to Jacksonville down the eastern seaboard of Florida. Any voice or data user on that eastern seaboard path can potentially access the system. I believe that the volume of voice and data remaining will not support a teleport in any of those markets served by the fiber. Further, if the Teleport does not exist in 1987–88, it will not have much chance of surviving due to the continued growth of fiber around the country.

The top 50 markets will get a teleport because of two factors still viable even in the presence of fiber: the presence of land and zoning restrictions on dish locations in the major markets and the large volume of corporate communications already existing in major markets.

Nevertheless, the medium-market teleport has a bright future. As stated earlier, to be successful, the medium-market teleport must be part of a group of shared management Teleports. This means that the teleport can have a basic overhead low enough to ensure that it can survive and prosper with a much lower requirement for business usage. In addition, these teleports will be a more cost effective method of delivery than fiber for a market with limited amounts of communications traffic. However, this is only a theory to be proved or disproved by time.

MARKETING TIPS

To be a financial success, the medium-market teleport must do three things: (1) Market; (2) market; (3) market. The question is how to market. In chapter 10, "Planning the Teleport," John Paul Rossie discusses the purpose and needs for planning. I suggest that you briefly review that chapter now. Let us identify the problems of marketing and some solutions that have worked for others.

Problem 1: No one understands what your teleport can do for their communications needs.

Problem 2: No one thinks the area needs a teleport.
With those two realities in mind, what should you do? Remember that your teleport provides a "Star Wars" type service. Below are some basic suggestions with notes as to how they apply to your marketing.

Do prepare a simple brochure of what communications problems your teleport can solve. The brochure could cater to continuing education of sales staff located in other cities. Keep the brochure simple and invite the prospect to develop questions while reading the brochure. The brochure should cause the prospect to be more receptive when a salesperson calls for an appointment.

Do speak about your teleport to every civic group in your market area. The people who make the decisions on communications belong to the Rotary, Optimist, Lions, and other clubs. Speaking before those groups is more effective than any ad you can buy. Prepare a simple 10 minute slide show that explains how satellite communications work and how your teleport interfaces with the industry. Be sure to include a list of current clients. They are a great source of influence.

Do join the local industrial development commission. This is normally an arm of the chamber of commerce. First, volunteer to serve on the communication committee. This committee contributes information about available communication services that become part of a package the industrial development commission mails to all prospective new businesses. Without your input, your teleport probably will not be mentioned when businesses are being approached to relocate in the area. Letting these potential new businesses know about the teleport can benefit you and the community.

Do compile a list of all companies in the area that make and distribute products. Prepare a letter to the companies' marketing managers. Ask for an appointment. Go and see them personally. Explain what you are doing, and ask how each manager introduces new products (videotapes, brochures, group meetings, etc.). Follow up with a detailed proposal showing how videoconferencing can save money.

Do contact each university or college in the market area. Locate the following individuals: continuing education director, media services manager, audio-video manager, department chairpersons, and any other person who may have information about which office might need wide area distribution. Set up an appointment with each person. Explore ways that they can use your services.

Do contact all the media (radio, TV, newspapers, and magazines) available in your area. Review the "media kits" of each, and try to purchase the media that makes sense. Use media buys to maintain identity in front of the prospective client. Don't try to sell services.

Do use direct mail for marketing. Target your letter to getting a face-to-face appointment with the person or persons responsible for buying the communications services of the prospect. Nothing can replace an on-target proposal made as a result of a face-to-face meeting.

CHOOSING TELEPORT SERVICES SALESPEOPLE

Where do you find teleport services salespeople? What skills do they need? The following answers are based on a particular management style and may not be usable in every medium-market teleport. First, whom should be hired? The salesperson need not be an expert in the satellite communications field. Initially, the biggest problem will be education. A good questioner who is a self-starter and very motivated can be more effective than a technical person. Such people are often in the insurance or securities industry. A careful training process will teach the salesperson what questions to ask so that a solid proposal can be constructed for a second appointment. At that second appointment, the salesperson can take a technical person along if it is necessary.

Compensation of these salespeople is a matter of personal choice, but I suggest a commission-based method.

What other skills does the salesperson need? I am a firm believer in a counselor roll, wherein the salesperson determines the needs of the prospect and then determines how and if the teleport can fill those needs. Encourage salespeople to tell the prospect when the teleport can not offer a viable service.

■   CONCLUSION   ■

The teleport, with its enhanced office space, fiber and microwave interconnects to the region, and access to all satellites offers valuable benefits to the medium market. Its physical location must be carefully selected in light of fiber construction. It requires a large staff, which could be shared between several teleports. Marketing will be more a process of education than selling.

For the real estate developer, a well-planned teleport can greatly enhance an office park.

Requiring careful planning, execution, and operation, the medium-market teleport represents the one unique opportunity for business community enhancement and profit that is available today.

Author

Edward A. Eagan
Managing General Partner
Central Florida Teleport
Ocala, Florida

Edward A. Eagan has an extensive background in the field of satellite communications and cable television. He was a founder of the Entertainment Sports Programming Network ("ESPN"); he was retained by Microdyne Corporation to work in the marketing of its satellite products; he formed a consulting firm, "Eagan & Associates," to provide up-to-date information and consulting services in the satellite communications field; he is Chairman of the Board of Wiresat Corporation, which was formed to build and acquire private cable television systems; and he formed and is developing the Central Florida Teleport, a satellite communications transmitting facility. Eagan has written numerous articles regarding satellite communications and is credited as a pioneer in developing the private cable television industry in its early stages.

Chapter 7

Ohio Teleport Corporation— Planning for a Dynamic Future

Miklos B. Korodi
The Ohio Teleport Corporation

Dr. Dennis K. Benson
Appropriate Solutions, Inc.*

Outline

---

* The authors gratefully acknowledge the help of Diane Bartosic, Director of Customer Relations, Ohio Teleport Corporation and Sandra Marks, Senior Partner, Appropriate Solutions, Inc. in critiquing and commenting on each of the innumerable drafts of this chapter.

END-USER CORPORATE PARTNERSHIP AND CONTINUING ADAPTABILITY

"Teleports: The Urban Alternative" appeared in *Computer Decisions* in April 1984. "Too Late for Teleports?" appeared in *Computer Decisions* on 7 May 1985. In only 13 months, how did teleports go from the state-of-the-art telecommunications service to a has-been?

Whether teleport executives are goats or heroes depends entirely on the concept of teleport they follow. The definition of a teleport must constantly evolve if it is to remain profitable and beneficial to users. Teleport executives locked into has-been definitions of their project need to update their resumes.

The conventional (i.e., dinosaur) definition of a teleport is a real estate development with sophisticted telecommunications capabilities. However, the real teleports of today and tomorrow must primarily be a set of ever-changing telecommunications services. These services cannot be limited by real estate property lines or by today's telecommunications technology.

Describing telecommunications technology is like trying to paint a wave: It doesn't hold still long enough. Those teleport projects with huge fixed assets may become operational, but in the end analysis these have already been eclipsed by advancing technology.

This tragically expensive obsolescence applies to the major telecommunications real estate projects as well as to the capital intensive fiber optics projects around the country.

The Ohio Teleport Corporation (OTC) is based on the principles of end user corporate partnership and continuing adaptability to evolving technology. What makes this concept so radically different and our confidence in its survivability so high?

A RATIONAL APPROACH

OTC is a custom service telecommunications company. Its purpose is simple: to facilitate the transmission of two-way data, audio, and video signals between businesses throughout Ohio, the United States, and the world. To accomplish this, OTC has designed a system that can appropriately integrate many different technologies—satellite links, microwave radio (digital as well as analog), coaxial cable, fiber optics, and terrestrial support systems.

OTC believes that it must plan to be able to respond to a wide range of ever-changing customer requirements—points served, equipment configurations, transmission speeds, and so on.

Company strategy has been influenced by five major factors that any successful teleport will have considered.

Deregulation

The deregulation of telecommunications is *the* precondition for the existence and (not inconsequentially) the potential profitability of teleports. The monopoly conditions AT&T enjoyed for so many decades are gone forever. An arena of entrepreneurial endeavor that has been closed for most of this century is now open.

Opportunities are legion, although the environment in which teleports operate is very unstable. There are still legal questions for which there are no answers, only guesses. That which can be deregulated can also be reregulated. Following the legal climate is critical, but it is often impossible to understand the legal situation.

A lot of people are making a lot of promises. The impact of deregulation is a proliferation of options, but unfortunately not an overabundance of quality. As more and more options open to business, the confusion increases.

Few people know what to do, how to evaluate competitors, how to choose between all the services that scream "I can do it cheaper." Even fewer have the knowledge to look beyond the "cheaper" to find the ones who could really do it better.

For the first time, because of deregulation, people have to think about their telecommunications. It used to be easy to rock in the cradle of Ma Bell and AT&T because it was the only nursery in town.

Computerization

Ohio Teleport is focusing initially on data communications. The scramble for voice traffic is both enormous and enormously confused. Video business requires a completely separate approach. Data communication, on the other hand, is an immediate specialized market which today has no substantive competition.

The demand for data services escalates daily. A study by Business Communications Company published in 1985 ("The Data Transmission Services Industry") pegs the 1983 revenue figure for data alone at $3.8 billion, with a forecasted rise to $11 billion by 1993. We believe the actual figure will be much higher.

Who is responsible for the surge in information transfer, and where should we target our efforts? The market for data communications is not merely the Fortune 500 companies with branches dotting the landscape. These companies clearly compose today's market, but tomorrow's market is the smaller user. This is where the exponential growth will occur. It is within the financial ability for virtually every corporation in the United States today to have communicating computer capability.

Every six months a new level in computerization is achieved. The future of the microprocessor in our lives is bounded only by imagination. This technology opens new vistas for small companies and completely new and lucrative businesses for those with imaginative and practical approaches to applications.

In Ohio 98 percent of all businesses have fewer than 100 employees. These companies cannot justify the high cost of 24-hour dedicated data circuits and may in fact have little need for it now. These companies will grow into steady customers. This is the future market for commercial data telecommunications.

Appropriate Technology

Service is at the center in the design of Ohio Teleport. Although teleports are commonly associated with satellite communications, there are many different ways to transfer signals from point A to points B, C, and D. To survive, teleports cannot afford to focus on single technologies for transmission of signals. To become capital intensive and focus on a single technology is to guarantee not only technological obsolescence but fiscal demise as well.

Sound concepts for teleports must incorporate a problem-solving orientation. Each customer will have unique conditions to meet. That is the same reason Cadillac dealers exist next door to Chevrolet dealers.

Much has been made of fiber optic technology and its great degree of error-free transmission. Microwave is doomed, say some, because of the huge capital-intensive fiber optic projects in the country.

If air transportation was all the country needed, interstate highways would be obsolete. Successful teleports will offer appropriate channels of transmission oriented specifically to the needs, requirements, and budgets of customers. Single transmission mode communication companies are candidates for acquisition, not investment growth.

Customer Relations

Where has AT&T failed? Our 1985 market study gave one answer: AT&T has few loyal friends.

All the years of monopoly lulled AT&T into an all-too-common public utility mentality: You don't need to be nice to your customers because they have no place else to go.

Within the "I can do it cheaper" scramble for audio communications customers, absolutely no climate of brand loyalty to these new telecommunications providers is being developed.

Look at the parallel situation in consumer banking. An increasing number of state and local financial institutions across the country are failing. Is this financial malfeasance? In some cases it is, but in many other cases, and providing a root condition for failure, these financial institutions have responded merely to a clamor for higher interest rates on savings.

The main way banks have been getting new business is by offering a slightly higher interest rate. The customers, perceiving no real competitive advantage in staying with one banking supplier, have responded by bouncing their accounts around, or have stayed with their institution because of simple inertia rather than conscious choice.

People will accept higher prices if the *value* of the service is worth the price paid. Most teleports promise lower rates and can deliver this; but if price is the only advantage, one telecommunication service will be replaced very quickly by the next offering a lower price.

A lower price is essential, but equally important is an excellent customer service program. People may try the service because of price but will keep the service, despite possible competition, because of value.

According to our research, the value that Ohio companies seek consists of:

Service reliability.
Transmission quality.
Monthly cost.
Redundancy.
Cost stability.

Service reliability was rated higher in importance over monthly cost by 34 percent of the companies surveyed. Critical to maintaining this service orientation is heavy involvement of users in management.

Uncertain Futures

We know two things about technological advancement: (1) It is geometric, and (2) it is unpredictable. Both statements are conservative.

Every teleport has its plan. Sometimes the word *plan* is in all capital letters and written in gilt in Old English script. Conservative accountants and financial advisors as well as certain investors expect a well-documented five-year plan. These plans look great, but none of them will ever be executed the way it is written. In only two years the realities on which those plans were based will completely change.

It is hard enough to look ahead six months with any confidence. It is extremely difficult to justify the time and expense to prepare detailed five-year plans that don't have a snowball's chance of being accurate.

That does not mean one should not plan. It is important to provide appropriate state-of-the-art telecommunications service to customers, since it benefits the customers to do so. However, this is a *philosophical,* not a *technological* commitment. The newest and sexiest technology is not always the most effective way to meet the customer's needs.

This appropriate technology may be a fiber optic link or two cans with a string. The only teleports that will grow and survive in an era in which the newest technology breaks in the journals before the last technology is in production are those that consciously adopt a philosophy of appropriate service.

That translates into flexibility. OTC realizes that something could happen tomorrow that may necessitate making changes in the service provided. Teleport managers have to realize that they are not solely in the business of providing passive communications links but may need to be involved in other services as well.

These ancillary services, developed in direct response to customers' needs, are necessary adjuncts to any full-service teleport.

ANCILLARY SERVICES

Teleconferencing

Home computers in the early 70s were a novelty, and the purview of technologically oriented hobbyists. Today teleconferencing is a novelty.

Teleports can provide the same service to teleconferencers as they can to data communicators—band width. There is a significant difference, though. Thousands of companies want transmission time for data communications, and these companies have specialists inside who understand computers and data. However, teleconferencing is a new technology and has outpaced the human factors; people simply don't know how to use it.

Television has only been with the mass public for slightly over three decades. Yet in this brief time, people have developed very definite views about how TV is supposed to work. A teleconference is not supposed to be the Academy Awards, but that is the expectation of many. Full-motion teleconferencing is and will in the foreseeable future remain a fad—a one-time conference, expensively staged and rehearsed.

For a teleport to receive significant video revenue, it has to become involved in education—teaching people how to use this developing technology.

Inexpensive video conferencing facilities will not sell video conferencing communications. Remember two-way cable? Technology completely outstripped practical application. The result was a lot of broken two-way cable promises, a situation that the cable industry is still reeling from today. Video-conferencing is dangerously in the same league here.

The future is not in full-motion video conferencing. The real future, the area in which major growth will occur, is audiographic and videographic conferencing.

Except for the occasional special cases, corporations will not use fancy full-motion studios that they have to cross town to use. Videographic conferencing should be in the conference room, not across town. This technique most closely parallels how people hold working meetings. When this type of teleconferencing is used daily for the real work of a company, it will become a meaningful profit center for any teleport.

Other Services

Teleports will also need to address other services in the future. If the evolving design of teleports is indeed market driven, then different teleports with different types of customers will find different needs.

Important to note here is that the growth of communications will be with smaller users. A teleport dependably providing a wide range of communications services will have an enormous entrée to these companies as they expand. Teleports that act as problem solvers rather than telephone poles will find themselves with a rather diverse portfolio of services.

These might include data storage, research consulting, planning consulting, education (seminars, workbooks, materials), video consulting (planning, programming, implementation), specialized equipment design and development, various backup services, and so on.

"But we've always done it this way!" is a one-way trip to bankruptcy. At best, following that philosophy will mean that such a teleport will totally fail to maximize its full income potential.

UNCERTAINTY PRINCIPLE

We have already said that teleports exist is an environment of excessive uncertainty. This must be acknowledged in the core of the organization. Operating in this environment requires massive tolerance for ambiguity, and success is determined as much by perseverance and serendipity as by anything else. No teleport can hope to succeed without planning for the unplannable. Five key factors affect planning.

Transmission Speed

Remember when dial up was new and we were astounded at 110 baud? Was it only 1984 when 300 baud was the standard? Was it just 1985 when 2,400 baud modems were becoming commonplace and within a specified radius a 9,600 baud modem could be used on an unconditioned phone line? We also have T–1 at 1,544,000 baud and fiber optics at 10 million baud.

Anyone who wants to guess what the standard will be six months from now would probably find the state lottery a safer bet.

Progressive teleports must be adaptive and avoid locking into technology that limits this adaptability. The state of the art is going to make time-shared, dedicated communications increasingly cost effective for the majority of American companies.

Transmission Channels and Destinations

We have T–1, microwave, satellite, coaxial cable, fiber optics, and so on. Any teleport that considers itself essentially a real estate development with an antenna farm is already a dinosaur.

We already have "smart" buildings wherein an internal cable system communicates to a microwave or fiber optic link to a satellite dish, which crosses the continent and reverses the process in a blink of the eye. A teleport is or should be involved in point-to-point and multipoint communications service. The extent and definition of this service varies by customer and will change as change is indicated.

Where the transmissions go is important. Getting the signal up has to be matched by getting the signal down again. Research shows that intrastate traffic is as important as interstate traffic, if not more so in some respects. International traffic is a sleeper. With automatic switching between satellites, the world becomes the footprint.

Teleports cannot build sound businesses by serving only high-volume users with satellite service to New York and Los Angeles. A full-service teleport has to be able to deal with point-to-point as well as point-to-multipoint throughout the country. It has to be as cost effective going across the street as it is going across the state or the country.

Growth

The areas of growth are predictable only in gross aggregate terms. The Fortune 500 is the tip of the market. Teleports must be alert to the growing small-company market. What makes this market so explosive is that as costs of communications technologies are decreasing, applications are expanding and companies are growing.

Finding the market is not a simple process of saying "customers need sales of $100 million to be prospects." Prospects are identified by need, not by size, and too many teleports are completely blind to this fact.

The technology will encircle the small companies whether they grow or not. Try writing a five-year plan using this as your basic assumption.

Partnerships

Three main factors make the Ohio Teleport Corporation development unique among its peers:

1. It is not real estate based.
2. It is full-service, point-to-point, and multipoint oriented.
3. Its founders are users.

All three factors combined, but driven by the third factor, mean that Ohio has progressed more rapidly than have other teleport programs. OTC's Board of Directors has a completely different view of the world. Too many teleport projects are trying to sell fancy new office buildings, which incidentally have sophisticated communications technology added on.

Companies won't move. While space has become more expensive, communications capability has become less expensive. Successful teleports maximize the customer's bottom line, and that does not mean requiring them to build new buildings and invest in a long-term, high-cost physical plant. Users know this, and real estate brokers are learning it. Teleports heavily oriented toward real property are building the Maginot Line of telecommunications. Service is operative, not physical plant.

THE FUTURE

Successful teleports in the future will have planned for technological discontinuities and will have a philosophical commitment to service.

Keep in mind what a teleport really is—a *tele*communications *port*al; it is a gateway for customers. However, the teleports of today will cease to be independent entities in the future. A good teleport's pragmatic approach to telecommunications service will become so overwhelmingly compelling, as a business and as a profit center, that in the coming years individual teleports will be absorbed by the next telecommunications conglomerate. We live in an era of aggressive corporate acquisition which will continue to produce some rather unusual bedfellows.

Authors

Miklos B. Korodi
President and CEO, The Ohio Teleport Corporation
Columbus, Ohio

Miklos B. Korodi is president and CEO of The Ohio Teleport Corporation, a full-service telecommunications company headquartered in Columbus. He joined Ohio Teleport in October 1984.

Skilled in guiding emerging companies from planning through development, Korodi helped establish Warner Amex Cable Communications, Inc. in 1979. In addition to full business responsibility for this company, he invented and patented the prototype Warner Amex Security System for two-way burglar, fire, and medical alarm services. Korodi was with Warner Amex corporate from 1977 until October 1984. He advanced from vice president and general manager of QUBE, the two-way cable system in Columbus, to senior vice president, new business development, prior to becoming president of Warner Amex Security Systems.

He holds seven patents on inventions, including a freight container bolster and side-loading container device that he developed for General American Transportation Corporation.

Active in community service, he is a member of the Rickenbacker Port Authority and Ohio Legal Services board of directors, and serves *ex officio* with the Columbus Area Cable Television Advisory Commission.

Korodi was born in Budapest, Hungary, in 1937, attended schools in Vienna, Austria, Budapest, and Melbourne, Australia, and studied electro-mechanical engineering.

Dennis K. Benson
President, Appropriate Solutions, Inc.
Columbus, Ohio

Dennis K. Benson is president of Appropriate Solutions, Inc. a research and management consulting firm based in Columbus, Ohio. Dr. Benson serves as Senior Research Consultant to the Ohio Teleport Corporation.

He is a founding partner of Appropriate Solutions, Inc., which was incorporated in 1978. The firm provides a wide range of services to both private and public sector organizations that have information needs or problems.

From 1973 to 1975 he was with the Academy for Contemporary Problems serving as Fellow of the Academy and director of The Benchmark Program, an experiment in urban social reporting and community-guided research. He received his doctorate in policy analysis from The Ohio State University in 1976 and was a free-lance consultant until 1978.

He has written numerous papers and monographs on such subjects as foreign policy behavior, social area analysis, voluntarism, human services information systems, and survey and research methodology.

He has consulted with such organizations as ABC News, Highlights for Children, Ohio State University, Ohio Department of Health, and Warner Amex Cable Communications. He is listed in *Who's Who in the Midwest* and *Who's Who in Business and Finance.*

Chapter 8

Bay Area Teleport—
A Case Study

Ronald H. Cowan

Bay Area Teleport

Outline

THE COMMUNITY

BAY AREA TELEPORT
  Earth Station Facility
  Electronic Nerve Center/Microwave
    Site

HARBOR BAY BUSINESS PARK

COMMUNICATING TELEPORT
  ADVANTAGES

FUTURE SUCCESS

Much has been written about the Information Age and the new era of information and communications technologies. The use of such phrases as *teleports, shared tenants services,* and *intelligent cities* has become common, although there are few examples in which these concepts have become a reality.

Harbor Bay is a new, master-planned community under development within the city of Alameda, California, that integrates a number of information and communication technologies, along with a variety of other elements, into a multidimensional "intelligent" real estate community. Master-planned communities afford planners the opportunity to anticipate advances in technologies and to provide flexibility for technologies to evolve within the planned community. Doric Development, Inc. is the managing general partner of Harbor Bay Isle Associates, which developed the Harbor Bay project. During the early stages of planning, Doric included the communications infrastructure, which later permitted the evolution of the project into a "teleport" and an "intelligent city"—even though the precise uses of the communications infrastructure and easements and the term *teleport* were then unknown.

THE COMMUNITY

Harbor Bay is a 917-acre "new community" project that occupies a bayfront crescent of land in the San Francisco Bay Area. The community includes: (1) Harbor Bay Business Park, planned to accommodate more than 5 million square feet of offices, research and development facilities, high technology, and support services; (2) the residential community of Harbor Bay Villages, currently zoned for 3,200 dwelling units (of which 1,600 are complete) and built to offer housing in a wide socioeconomic spectrum; (3) Harbor Bay Landing, a 10-acre commercial center with 87,000 square feet, including a major supermarket, shops, restaurants, and financial facilities; and (4) recreational facilities including Harbor Bay Club with its clubhouse, 20 lighted tennis courts, pool, racquetball courts, exercise studios, weight training center, and fitness evaluation facility.

The communications infrastructure built into the Harbor Bay Community provided the foundation for creating two separate companies to integrate a variety of information technologies into the Harbor Bay real estate project. The companies, Bay Area Teleport (BAT) and Harbor Bay Telecommunications (HBT), are each equally owned by Doric Development, Inc. and Pacific Telecom, Inc. (PTI), a majority-owned subsidiary of Pacific Corporation. PTI, based in Vancouver, Washington, is the sixth largest non-Bell telephone company in the United States and is the owner of Alascom, the primary long-distance carrier for Alaska. PTI, through its subsidiaries and partnerships, is involved in virtually all aspects of the communications industry. Doric and PTI have committed many millions of dollars in capital to HBT and BAT to develop what many believe are unexcelled communications services for tenants at the Harbor Bay Business Park.

Harbor Bay Business Park is one of the first projects that qualifies as a teleport within the definition of the term adopted by the World Teleport Association, an international organization of public and private entities associated by their interest in teleports and intelligent cities. The World Teleport Association defines teleport to require the presence of three elements: (1) satellite access; (2) a regional telecommunications distribution system; and (3) connectivity to a significant real estate development.

BAY AREA TELEPORT

BAT provides the Harbor Bay real estate development with satellite access and a regional telecommunications distribution system and, thus, qualifies Harbor Bay Business Park as a teleport.

Earth Station Facility

BAT's satellite access originates from its earth station facility at Niles Canyon, located on a 347-acre site in Alameda County, nine miles from the BAT control center at Harbor Bay Business Park. The facility is capable of supporting up to 36 satellite earth stations, with frequency coordination for full band, full arc operation. The location of the site is within a narrow geographical band on the West Coast in which every North American domestic satellite as well as both Atlantic and Pacific Intelsat satellites are visible. The facility offers access to all of these satellites with no RF interference.

The earth station facility located at Niles Canyon will include a common service building, uninterruptible power supply, 24 hour on-site maintenance, security, remote alarm monitoring, and a full menu of earth station design, engineering, and construction services. The complex is accessed by the BAT distribution system, with plans for secondary fiber optic access to provide redundancy. BAT leases space at the facility to a variety of satellite carriers that will install antennas at the facility. Satellite carriers leasing space at the earth station facility simplify the difficult and time-consuming tasks associated with establishing earth stations and reduce the period of time it takes to become functional merely to the time it takes to order and install communications equipment. The tasks of locating a site able to view the appropriate satellite without outside interference, obtaining zoning, getting frequency clearance and other governmental permits, physically developing the site, and connecting the facility to population centers have all been accomplished by BAT. As aesthetic and environmental concerns over earth stations increase, areas available for communications system installations will decrease, and BAT's earth station facility at Niles Canyon will become even more attractive to satellite carriers.

The San Francisco Bay Area Teleport, headquartered in Harbor Bay Business Park, Alameda, uses a unique digital microwave network to distribute communications traffic throughout the San Francisco–Oakland basin, and connects via fiber and microwave to a satellite earth station facility that is the largest shared use teleport in the world. *Courtesy: Bay Area Teleport*

Electronic Nerve Center/Microwave Site

BAT's electronic nerve center and central microwave site are located at the Harbor Bay Business Park situated on a 135-foot tower designed by the world renowned architectural firm of Skidmore, Owings & Merrill. The earth station complex is connected to the hub facility at the Business Park by a BAT microwave link. BAT's microwave system is completely digital, with 15 primary nodes in nine Bay Area counties and in the Sacramento area. The system is designed for a data capacity ranging from 1.554 to 135 megabits per second. High reliability is built in with proven technologies and redundant equipment located throughout the system. Digital information will be carried via microwave antennas linked to businesses in ten Bay Area

FIGURE 8–1
Bay Area Teleport Microwave Network

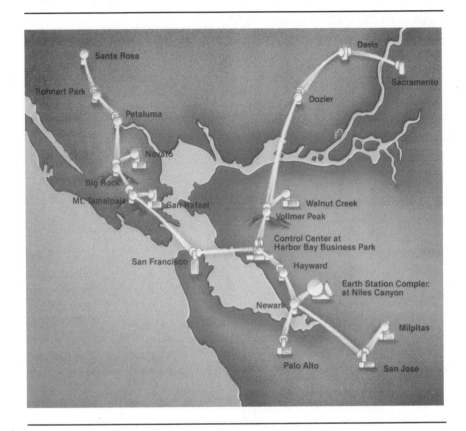

Courtesy: Bay Area Teleport

counties, including Harbor Bay Business Park in Alameda County and BAT's earth station facility at Niles Canyon.

The BAT microwave network provides communications access for nearly the entire Northern California area. The system extends over 3,600 square miles, initially from San Jose in the Santa Clara Valley in the south through San Francisco to Santa Rosa, 50 miles north of San Francisco. The first phase also incorporates the Sacramento/San Joaquin delta area east to Sacramento. Potential expansion plans include areas north and south in the Sacramento and San Joaquin valleys. Ultimately, there could be nearly 10,000 square miles covered by the microwave network. Access would then be available to a geographical area that includes every major university and state and junior college in Northern California, the hub of state government, California farming regions, major transportation centers, and the banking and commercial centers of San Francisco.

HARBOR BAY BUSINESS PARK

At Harbor Bay Business Park, HBT provides occupants a variety of shared tenant communications products and services.

HBT utilizes a multimillion-dollar, state-of-the-art, PBX-type InteCom IBX switch as the basis of its advanced digital communications system. By sharing the resources of a large and sophisticated switch, HBT subscribers have access to myriad capabilities and avoid the risk of equipment obsolescence. Reduced capital expenditures and extremely attractive rates result in substantial business communications cost savings. The features and services available include:

Enhanced, integrated voice/data switching.

Sophisticated electronic security and surveillance system.

Full-motion video.

Shared word and data conferencing.

Message center.

Consulting and systems design.

Synchronous and asynchronous data transmission.

Fire alarm monitoring.

Modem pooling.

Low-cost routing of long-distance calls.

Office automation and office services.

Call detail recording.

Shared word and data processing.

Facsimile.

Paging.

Billing account codes.

Voice mail.

Network management and design.

Automatic call distribution.

Data services network access.

System components are fully redundant and operate on a full-hot standby basis. The system is equipped with online diagnostic capability and is monitored around the clock by HBT personnel. Spare parts are kept in inventory, also on-site.

Subscribers, for the most part, have the option to alter or terminate any service at any time and may upgrade or change hardware and software at no financial penalty. To date, businesses locating at the Harbor Bay Business Park have been unanimous in seizing the competitive advantages offered by HBT.

The communications infrastructure built into the Harbor Bay project from its earliest planning stages has enabled the project to take advantage of the latest innovations in information technology. Efficiency in collecting, recording, and distributing information generated by today's businesses is becoming harder to achieve, in part, because the quantity of information is expanding almost geometrically. The communications costs of many companies are for the first time nearing or exceeding their facilities costs. This trend will continue as the 21st century approaches and the predictions that the Information Age is upon us are realized.

COMMUNICATING TELEPORT ADVANTAGES

Those business people who recognize the trend and consider the information and communications consequences of a location decision will be able to take advantage of the efficiencies inherent in locating facilities at a teleport or within an intelligent city. To assist those who make facilities location decisions, marketing representatives must illustrate the communications efficiencies that can be realized at such teleports as Harbor Bay. The illustration of such efficiencies is technical. Thus marketing representatives of communications entities must take an active part in all stages of contact with potential occupants of the real estate development. This observation was confirmed by recent Urban Land Institute research, wherein it was concluded that successful shared tenant service projects have two things in common: (1) shared tenant services marketing begins simultaneously with leasing; and (2) the marketing effort requires close cooperation between the real estate developer and the communications service provider.

At Harbor Bay, management and the marketing staffs of both telecommunications and real estate work closely together in all phases: planning and strategy, preparation of proposals and support materials, and structuring and negotiating a proposed transaction. In addition to the facilities department, marketing efforts are targeted at the data processing group, financial department, telecommunications group, and executive committee—in an attempt to emphasize information and communications requirements and opportunities as part of facilities location decisions. Prospective tenants are encouraged to think of a relocation decision as a strategic decision that affects not only their location and the price to be paid for space, but also the impact the decision has upon data services and all aspects of communications.

A 1985 Urban Land Institute survey closed its chapter on Harbor Bay Business Park with this comment: "The Harbor Bay project is clearly one of the more ambitious undertakings in the technology-enhanced real estate business. The developers have set their sights high; they are spending a lot of money to make sure the technology is done right, and the service is of high quality."

FUTURE SUCCESS

The location of businesses in teleports and intelligent cities will be commonplace over the next decades. Businesses will appreciate the efficiencies that accrue from combining the advantages of locating in a well-conceived and high-quality business community, such as Harbor Bay Business Park. They will appreciate the advantages of communications and information distribution systems, such as those offered by BAT and HBT. It is only the farsighted, however, who anticipated these trends and planned to take advantage of them, who will participate in the first phase of the "information revolution." Harbor Bay Business Park, having the three characteristics of a teleport plus a menu of shared tenant communications services, is uniquely situated to take advantage of the evolution into the Information Age.

Author

Ronald H. Cowan
Bay Area Teleport
Alameda, California

Ronald H. Cowan is president and chief executive officer of Doric Development, Inc., and chairman of Harbor Bay Isle Associates. His projects are the recipients of 17 local, state, and national awards for excellence. In 1972 he formed Harbor Bay Isle Associates in partnership with Utah International Inc. and directed the planning and development of Harbor Bay as a self-contained community.

Cowan led former California Governor Jerry Brown's first trade mission to Japan. He has served on the Economic Development Commission and the Industrial Innovation Commission for the State of California and currently serves as a Director of the World Teleport Association.

He is a member of the Urban Land Institute, a director of the Oakland, California, Chamber of Commerce, and a past director of the Oakland International Host Committee. He is active in a number of other community and industry organizations.

Chapter 9

Teleconferencing, Teleports, and Intelligent Buildings

Thomas B. Cross

Cross Information Company
Intelligent Buildings Corporation

Outline

Source: Adapted from Thomas B. Cross and Marjorie Raizman, *Telecommuting: Work Strategies for the Information Organization* (Homewood, Ill.: Dow Jones-Irwin, 1986) © Dow Jones-Irwin, 1986.

The emergence of intelligent buildings and teleports will drive building development in the late 80s. Teleconferencing is expected to become an important services offering to tenants, and accordingly, will create a demand for telecommunications services that can be provided by teleports. This chapter will discuss the expected growth in teleconferencing and the resulting impact on buildings and teleports.

Building developers face both soaring energy costs and more sophisticated and demanding tenants. Those who do not adopt the new technologies inherent to an intelligent building may find themselves locked out of an increasingly competitive market. According to R. Craig Blackman, director of the telecommunications program at Barry University, "Intelligent buildings, quite simply, will be the competitive edge in higher tenant occupancies and profits. Many will ignore this great opportunity, but this is a significant trend that will not go away." (Personal interview with the author, December 3, 1985.)

Intelligent buildings propose to link all operations through a central, or host, computer communications system. Thus all environmental control, energy, lighting, fire, safety, and security systems will be centrally controlled to the extent that they will almost run themselves. Intelligent buildings, in many cases, will be connected together or integrated with teleports that support telecommunications, data communications, and other existing or future information technologies. Having completed an in-depth market research study on intelligent buildings, Cross Information Company (CIC) foresees that most office buildings (large or small) will soon provide a wide range of enhanced or intelligent services and, in some cases, teleports. Stemming from the above emerging forces, the reasons for this growth in telecommunications sources follow:

- Tenants have a growing awareness of their ability to achieve more cost-effective rent. This does not mean lower rent payments but, rather, a fundamental shift from paying separately for office services to combining those services (telephones, teleconferencing, data/word processing, etc.).

- There are growing competitive pressures from other developers providing these services.

- Technological innovations now permit combination of many formerly different services, such as electronic mail, networking, and data communications.

- Improvements in real estate and building management systems now enable many separately located complexes to be centrally managed.

Technological innovations in computer and telecommunications technologies have been the principal forces behind teleports and intelligent buildings.

The development of an "intelligent building" is no longer in question: Simple timing is the issue now. Onsite or interconnect teleports are not far behind. To illustrate the fast-paced growth of this technology, 25 years ago most offices had individual window air-conditioning units. These have since been replaced by centralized, multi-tenant systems, which give better and even personalized service at a lower unit cost. The intelligent building is emerging along these same lines, though in a more evolutionary than revolutionary process.

In designing teleports and associated intelligent buildings, it is essential to understand these new technologies as well as review the management styles and communications infrastructure of organizations that will use intelligent buildings. Intelligent buildings require similar set designs to integrate new technologies into the structure of the tenant organization. This reflects the fact that information technology, especially office automation, is likely to change the way in which organizations make decisions on the allocation and types of spaces needed.

Another reason for a comprehensive planning approach is that not all organizations benefit equally from the multi-tenant system. Developers must identify prospective tenants of an intelligent building and plan for their needs. Intelligent buildings are, for the most part, geared toward multi-tenant services. Large companies that can afford their own intelligent building systems and potentially their own teleport are reluctant to get locked into a building system that either duplicates what they already have or appears to be less reliable than their own equipment.

On the other hand, medium-size and smaller companies benefit from an in-house system and teleport provided by a building developer. Smaller companies often cannot justify the cost of owning their own equipment but can lease high tech services just as they do office furniture. The actual cost to the tenant depends upon the choice of services and length of time the user is on the system.

Among the wide array of service offerings, teleconferencing will be one provided at the next level of intelligent building. According to Paul C. Daubitz, president of ATI TeleManagement of Boston,

> Teleconferencing is very "network-intensive," which translates into higher revenues for the owner/manager. The key will be in selecting the best teleconferencing system, training users, and marketing the service not just internally, but to the surrounding geographic area serviced by a teleport.

INTRODUCTION AND OVERVIEW OF TELECONFERENCING

Teleconferencing is a system that enables two or more people at multiple locations to communicate electronically. Users can exchange information and examine drawings, plans, or sketches without interrupting their work schedules or paying for costly travel. In practice, teleconferencing participants have found that such systems of communication can:

■ Greatly reduce the need for travel, thereby saving both time and expense.

■ Break the habit of meetings, thus encouraging conferencing to accomplish much more in less time.

■ Allow people to deal with issues as they arise, instead of postponing them until the next scheduled meeting. This increases the speed of decision making and improves organizational efficiency.

■ Increase efficiency by shortening meeting times.

■ Revitalize organizational effectiveness by providing a convenient form of communications.

■ Improve the quality and quantity of feedback from the field to headquarters.

■ Increase organizational loyalty and enhance job satisfaction by allowing many more individuals to participate in decision making.

Worldwide, management spent hundreds of millions of dollars to improve factory productivity in the 1980s but virtually ignored the office environment. As a result of this oversight, the cost of running offices steadily consumed larger shares of companies' operating budgets. Three factors have made improving the office environment important: (1) the skyrocketing costs of air travel, hotel and motel accommodations, management salaries, and car rentals; (2) the increasing importance of reducing information float or decision-making cycles; and (3) new technology that makes it practical to communicate through teleconferencing.

Since communicating is the most expensive element in operating an office, teleconferencing has become increasingly important as a supplement to corporate communication systems. Communications includes everything from answering telephones and posting notices on bulletin boards to writing letters and participating in meetings and conferences.

According to comprehensive management studies, executives and managers spend 75 percent of their time communicating. That is, more than six hours each workday is spent attending meetings, reporting on those meetings, and talking on the phone. Only about 22 percent of an executive's or manager's time is spent doing desk work.

Moreover, recent studies have revealed that 60 percent of all communications do not require face-to-face meetings. Topics discussed in meetings can often be handled in less costly ways. In fact, once most meetings and conferences are stripped of small tasks and formalities, their useful duration is usually less than half an hour. One third of these meetings seem to be for the sole purpose of exchanging information and not for decision making. (See Figure 9–1.) Teleconferencing is a viable supplement rather than a re-

FIGURE 9–1

Present Manager's Distribution of Time
Communicating information transactions:
    Meetings.   30%
    Telephone.   20%
    Travel.   20%
Seeking information transactions:
    Desk work.   30%

Future Manager's Distribution of Time
Communications/information transactions:
    People interfaces—Meetings, presentations, audio teleconferencing, video tele-
    conferencing.   40%
Travel: 10%
Seeking information transactions—system interfaces: 50%
    Dictation.
    Telephone/voice mailbox.
    Computer teleconferencing.
    Viewdata—data systems.
    Decision support systems—assisted "thinking."
    Computer-assisted retrieval.

Present Nonmanager
Inputting information:
    Dictation, typing, data entry, and proofreading.   20%
Interpreting information:
    Telephone, mail.   35%
    Away from desk.   30%
    Waiting for work.   10%
    Absent.   5%

Future Nonmanager
Information analysis: 10%
Information transactions: 50%
    Project research.
    Meeting coordination.
    Arranging travel.
    Budget tracking.
    Purchase order tracking.
    Researching.
    Teleconferencing.
Administration coordination/support for meetings:
    Telephone, mail.   35%
    Absent.   5%

Courtesy: Cross Information Company

placement for face-to-face meetings because it allows an easy exchange of information without expensive, time-wasting formalities of traditional corporate meetings.

Factors Driving Teleconferencing

Corporate needs for teleconferencing are as varied in today's high-cost, information-based environment as are the media and applications developed to meet them. Generally, teleconferencing meets the need for:

- Frequent communication between remote sites.
- Communication across functional lines.
- Business meetings in generally inaccessible locations.
- Avoidance of high travel or telephone communications costs.

Despite early skepticism of those unfamiliar with teleconferencing, the overall experience of people who have actually tried such systems has been encouraging. Participants agree that teleconferencing has a very positive impact on the work environment and *management velocity*—a term coined to describe the speed and effectiveness with which problems are solved. Due to increasing costs of traditional forms of organizational communications, advances in technology, and the benefits of teleconferencing, executives and company owners turn increasingly to teleconferencing for speedier and more effective communication.

SURVEY OF TELECONFERENCING SYSTEMS

The following section is an overview of the different forms of teleconferencing presently available:

- Radio.
- Video.
- Slow scan.
- Audio.
- Audiographic.
- Visual-graphic.
- Computer.

Radio Teleconferencing

Because applications of radio teleconferencing are limited, this concept will not be discussed here beyond stating the following key characteristics of radio-based or over-the-air teleconferencing:

- It is a natural transmission medium used over widely dispersed areas.
- It is comparatively low cost compared with stringing wire.
- It is not often used when security is required.
- It requires special high-frequency receivers.
- It can be used as a paging system.
- It can be connected to the telephone network.

Video Teleconferencing

In using full-motion video teleconferencing, people at two or more different locations have contact almost as if they were seated in the same room. The four broad types of video teleconferencing are:

1. *Ad hoc*—The use of broadcast-type facilities as mobile uplinks; these are typically located in television stations and hotels. The objective is to reach thousands of people for a one-time shot, at a cost of between $14 and $200 per meeting. See Figure 9–2 on p. 156.
2. *Interstate corporate teleconferencing*—The use of private or public rooms located on the customer's premise or in a metro area. This is typically used for 6 to 10 people per location and costs $2 to $8,000 per meeting, depending on duration and distance. See Figure 9–3 on p. 157.
3. *Intrastate local/campus*—A set of privately owned rooms used for daily business meetings or education within a corporate campus or local community.
4. *International teleconferencing*—Incorporates both ad hoc and interstate conferencing as an evaluation phase. There are significant implementation issues, which are perceived as having high corporate market value. These are some of the key features of video teleconferencing:

- Delivers real-time video images.
- Allows for personal presence.
- Imparts body language and immediate emotions.
- Allows for rapid decision making.
- Is ideal for group as opposed to one-to-one discussions.
- Allows for a "crisis meeting" to take place quickly.
- Allows for "high impact," Hollywood-style events.
- Often requires large capital commitment.
- Requires large, on-going overhead and maintenance.

Problems associated with video teleconferencing include the high degree of system complexity and the small number of systems suppliers—both of which may result in comparatively high implementation costs. Much of the

FIGURE 9–2
Satellite Teleconferencing

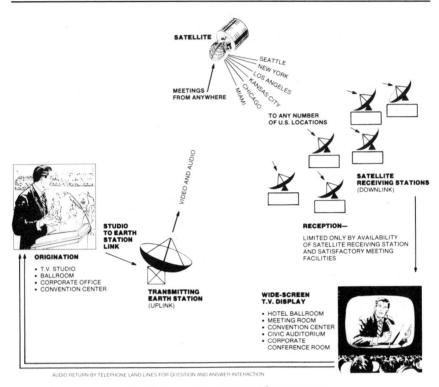

VIDEOSTAR TELE-MEETING® NETWORK

Courtesy: Videostar

teleconferencing industry is segmented into companies that provide a single component, such as audio speakerphones, television cameras, transmission systems, and so forth. A few entities, however, are emerging as total "systems suppliers." They use approaches that will bring down the corporation's costs of video teleconferencing. Approaches used to market video teleconferencing include:

- *Dedicated systems*—Teleconferencing rooms for the exclusive use of a single corporation.
- *Semidedicated systems*—Teleconferencing rooms primarily used by a single corporation, but available to others at certain times of day.

FIGURE 9–3

Courtesy: American Satellite Corporation

■ *Shared systems*—Teleconferencing rooms located in office parks, hotels, and so on, available to users on a reservation basis.

Although video teleconferencing is a valuable business tool, it can be prohibitively expensive if acquired on an individual tenant basis. Operated on a reservation or time-available basis, the shared center for tenants is one of the most popular applications of video teleconferencing because it permits the sharing of costs and benefits.

Slow-Scan Television

Slow-scan video, also called freeze frame, slide-show, or still-frame video, is much like a slide show. Slow-scan transmission is of high quality. "Still motion" images are sent every 8 to 35 seconds, depending on the bandwidth of the circuit (e.g., normal telephone lines average 35 seconds per frame). In most slow-scan systems, images are built upon the screen from left to right, or top to bottom.

The cost of slow-scan systems is much lower than that of full-motion video. Although full-motion video provides one of the most natural environments for conducting meetings, the still image is also used for person-to-person communications wherein body language is neither important nor desired. This is especially true, for example, in transmitting the printed page

TABLE 9–1
Trends in Video Teleconferencing

Bit Rates Required for Motion Video

1980	90.000	Megabits
1981	30.000	Megabits
1982	1.544	Megabits*
1983	0.784	Megabits
1984	0.056	Megabits
1987	0.019	Megabits
1990	0.009	Megabits

Cost of Video Teleconferencing Technology per Location

1980	$500,000
1981	$300,000
1982	$200,000
1983	$150,000
1984	$ 75,000
1985	$ 50,000
1987	$ 25,000
1990	$ 10,000

Video Teleconferencing Scale

1980	VLSI—Very Large Scale Integration
1981	LSI—Large Scale
1982	MSI—Medium Scale
1983	SSI—Small Scale
1984	RI—Roll about
1985	SF—Store Forward
1987	P—Portable
1990	W—Walk about

* See Figure 9–4.
Courtesy: Cross Information Company

or photography or when a speaker elaborates on a theme through the use of slides, view-graphs, flip charts, chalk drawings, magazines, or other non-moving material. See Figure 9–5. It is seldom necessary to introduce actual moving imagery in the form of film clips or videotape recordings, according to Glen Southworth, president of Colorado Video, Boulder, Colorado. See Figure 9–6. Slow scan provides the same type presentation in the teleconferencing framework. The features of slow-scan teleconferencing include:

- Uses normal telephone lines rather than broadband circuits.
- Provides portability and usability at remote sites.
- Provides nonmoving pictures for presentations similar to a 35mm slide show.

FIGURE 9–4
DS–1 Channel Expansion Multiplexer

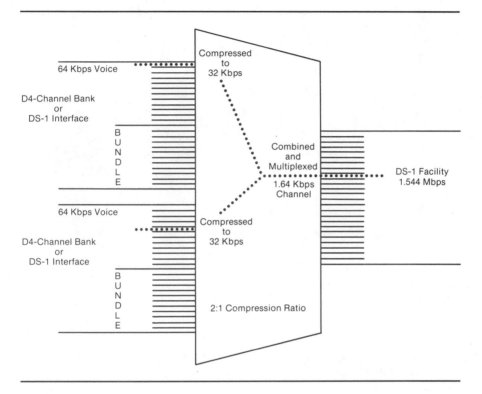

Graphic courtesy AT&T Information Systems

- Provides pictures or presentations that can be recorded on an audio cassette recorder or computer disk for later use.
- Makes available high- and low-resolution systems that take from two to more than seven minutes per image to transmit.
- Costs very little compared with full-motion video.

In the case of slow-scan teleconferencing, the single-picture transmission provides enormous savings in channel capacity when compared with conventional video use. A variety of slow-scan television picture transmissions have been developed over the past two decades. The present trend is to use conventional video cameras, monitors, and other system components in conjunction with "scan conversion" devices. These reduce the bandwidth of the television camera's output from a nominal 5 mHz (hertz is one cycle, mega = 1,000,000) to approximately 1 kHz (kilo = 1,000) for transmission over voice grade

FIGURE 9–5

Courtesy: Colorado Video

or normal telephone lines and a compression ratio of 5,000:1. This ratio is achieved by stretching out signal time from 30 pictures per second to 1 picture per 7+ seconds and/or reducing visual resolution. Because telephone line bandwidth is expensive, such bandwidth ratios represent large cost savings in transmission.

IBM's slow-scan video teleconference system presently comprises more than 110 rooms worldwide, with an additional 150 to be added in 1985. Using the IBM System, it is possible to transmit voice, black-and-white, or color video images of people and graphics, as well as high-speed facsimile and electronic storage of documents for recall and later transmission. Security arrangements allow for presentation and discussion of confidential IBM information. According to IBM, "[Teleconferencing] helps you solve problems faster by improving your access to people and information."

Slow-scan video has a number of applications. Some of these include:

- Teleteaching.
- Engineering design meetings.
- Telepublishing of news and information.
- Telemedicine: remote diagnostics, including X-ray analysis.
- Environmental monitoring and security.

FIGURE 9–6
Model 950 Teleconferencing System

Courtesy: Colorado Video, Inc.

When these applications have had more publicity, the use of slow-scan video will surely increase.

Audio Teleconferencing

Audio teleconferences can be assembled in one of two ways: (1) by dialing-out from the audio bridge or (2) by dialing in to a "meet me" bridge. In the dial-out process, provided by a number of companies, a conference operator or controller calls all participants sequentially before the meeting then bridges them together until all are assembled. A dial-out meeting can take up to 30 minutes to set up.

When using the dial-in or "meet me" bridge, often provided by a service bureau, the individual teleconferees call themselves into the bridge at a specified time. See Figure 9–7. Using a "meet me" bridge can reduce audio teleconference setup time by eliminating major delays, such as incorrect telephone

FIGURE 9–7

Courtesy: Confertech International

numbers, unavailability to take calls, and poor connections. A good speaker-phone is essential to successful audio teleconferencing. See Figure 9–8.

An audio bridge is a device that generally allows up to 48 (or more, depending on which system is used) people to be connected simultaneously. Telephone lines go into the bridge, where they are bridge amplified. Bridge amplification means that all participants have the same voice quality in the conference, as all are brought to the same sound level, and noise is filtered out. An operator controls the bridge console either by calling teleconference participants or by arranging a predetermined time for the teleconference.

Audio bridges are now being automated. Enhanced technology can provide the capability of holding daily multipoint teleconferences without the need to interface with a conference operator. A microprocessor coupled with a voice synthesizer gives greetings and instructions to the teleconference participants. As participants dial into the system, they are greeted individually by the voice synthesizer and told to introduce themselves into the teleconference. Access to the system is via touch or push-button tones generated by a tone dial pad on the user's telephone.

Operational features of an audio bridge can include:

FIGURE 9–8

Courtesy: Teleconferencing Systems International

Forty-eight port teleconference capacity.

Microprocessor controlled with automatic answer and instructions by voice synthesizer.

Access by off-premise tone dial key pad.

Software allowing subdivision of teleconference into simultaneous sub-teleconferences.

Automatic gain amplification compensating for network losses to provide high-quality audio to each participant.

Security features via touch pad, which lock out undesired, would-be participants.

Software customized to user's operational requirements.

Monitor jacks that provide "operator" interface, if desired.

Individual status lights on each port module to allow for talker identification.

Expander jack to allow linking of two systems for full 48-port configuration.

Twenty-four-hour-a-day access from any telephone in the world network.

Compatibility with all types of amplified speakerphones or teleconference systems.

The primary issues in selecting an audio bridge include:

Speech quality.

Ability to interrupt.

Number of concurrent conferences.

Size of conference.

Ease of setup speed.

Ease of operator training.

Security.

Graphic terminal compatibility.

Transmission of data communications or slow scan.

Unattended operator capability.

Technology—analog or digital.

Preparation for the meeting is vital to achieving a successful audio teleconference. Some of the key components are:

Installation and testing of equipment.

Schedule of program dates.

Deadlines set for handouts and audiovisual material(s).

Instructor(s) contacted/trained regarding their role(s).

Coordinators selected for each participating site.

Guidelines given instructors(s); these include information on participants, site location, and names of coordinators.

Coordinators given handout/training explaining their roles/duties.

Practice run of program.

Characteristics of audio teleconferencing include the following:

Greetings, introductions, and farewells must be incorporated into the teleconference itself, since arrivals and departures are invisible.

Participants may remain relatively anonymous.

Active listening is required if conferees are to be involved in a meaningful way.

Vocabulary and syntax have a heightened impact on listeners, hence on the conference's effectiveness.

Vocal characteristics, such as inflection, volume, and speed, have magnified impact on the communication exchange.

Pauses in conversations and long silences have their own value, which can be positive or negative.

There may be difficulty in sending or receiving messages that depend on visual cues, such as eye movements or body language.

See Figure 9–9 on p. 166.

With adequate preparation, audio teleconferencing can be a powerful and inexpensive communications tool. The following are some of its more viable applications:

- Coordination of internal administrative affairs.
- Interface with field offices.
- Product releases and telemarketing.
- Training of field personnel.
- Coordination of remote manufacturing operations.
- Public relations and product development.

Audiographic Teleconferencing

Audiographic teleconferencing usually adds visuals to voice for increased communication effectiveness. This mode of teleconferencing is distinguished by the various types of equipment used. Key features include:

FIGURE 9–9

Courtesy: Virginia Ostendorf, Inc.

- Graphics that reinforce audio teleconferencing by enabling design, engineering, and editorial collaborations to occur "live."
- Visual awareness of the participants, relevant documents, and a setting's ambiance.
- Graphics that allow relevant documents to be conveyed during meetings.

There are various types of audiographic teleconferencing:

- Telewriters which are multicolored electronic pens. Writing is done on the video terminal, and the system has an integrated workstation and speakerphone.
- Electronic chalk and marker boards.
- Personal computers, including (1) the personal image computer with its facsimile devices, (2) lap boards for freehand drawing, (3) mouse input devices.

See Figure 9–10.

FIGURE 9–10

Courtesy: AT&T

The addition of graphics to audio teleconferencing permits specialized applications of teleconferencing, such as descriptions of:

Engineering designs.
Building layouts.
Advertising.
Interactive equations.
Legal briefings.

Visual-Graphic Teleconferencing

A further evolution of the audiographics concept in electronic teleconferencing is visual graphics, computer graphics-enhanced video teleconferencing using freeze-frame video. The freeze-frame technique adds the dimension of video, and for an additional cost, audio support. Synchronization of these media incorporate their separate advantages into one system. Visual graphics clearly make teleconferencing effective at a fraction of the cost of a full-motion video setup.

This type of system is exemplified by the use of computer graphics software to enhance freeze-frame video teleconferencing. In this arrangement, graphics are created on a personal computer and uploaded to the mainframe host. The host computer carries graphics, video, and room-lighting control commands over a data network to those remote sites participating in the point-to-multipoint (one sending location to many receiving locations) video conference. The central computer can bridge software links up to 10 sites at either 9.6 or 19.2 kilobits per second. The full-color, freeze-frame video operating at 56 kilobits per second costs about one-10th as much to transmit as does a full-color, full-motion video transmitted at 1.5 megabits.

The advantages of visual-graphic teleconferencing are:

- The emphasis on moving graphics and data rather than moving individual people.

- Spreadsheet software, featuring columns, charts, graphs, and text that allows for system application in many areas.

- When compared with full-motion video teleconferencing, software programming on a personal computer (PC) is a substantially faster, simpler, and cheaper process than a production crew shooting on film or tape. In addition, editing involves inserting a floppy disk into the PC drive and executing desired changes, whereas film and video editing is a slower process.

- This type of system can be hooked up to the large numbers of mainframes and personal computers already being used by many organizations. By stressing computer graphics over video, these companies can use their existing data processing resources to enhance the cost-effective video teleconferencing. The proliferation of PCs has driven and altered data and voice communications through teleconferencing. In fact, this video breakthrough will probably make personal computers the dominant and creative force in teleconferencing.

This system's orientation as a data medium rather than as a video showcase becomes apparent upon examining the video teleconferencing room. Since graphics are transmitted before the teleconference, a projection screen shows vivid computer images while freeze-frame images are watched on an adjacent TV monitor. A floppy disk is loaded into one of the PCs in the room. As the mainframe sends data over a proprietary network, a checklist appears on the projection screen. In about five minutes, the software data is received at up to 10 remote sites. The mainframe manages graphics, video, and controller commands from this one central location.

The applications of visual-graphic teleconferencing embrace all previously mentioned forms. Therefore, it is an especially effective tool for entities with comprehensive communications needs.

Computer Teleconferencing

Based on the written form of communication, rather than video or voice, computer teleconferencing (CT) engenders social interaction that differs from the audio, audiographic, or video modes.

Often called computer conferencing, computer-aided communications (CAC), or computer-aided networking (CAN)—CT allows people in different geographic locations to conduct ongoing meetings using video terminals (CRT–VDU), personal computers, or computer systems. An electronic message system records each participants' input. Any conferencee may access, read, and respond to these communications, regardless of whether or not other participants are communicating simultaneously. The system thus provides a verbatim log of the meeting.

The major attraction of computer teleconferencing is that, unlike other forms, all participants do not have to be at their desks at the same time. The computer's capability to provide asynchronous, nonreal-time, or store-forward communications offers participants extraordinary flexibility. Computer teleconferencing is not dependent on any one person's time schedule, nor the variables of rain, sleet, sickness, distance, travel, and time zone differences. This technique has proven to be highly effective for managing ongoing project activities. The bottom line is that computer teleconferencing prevents frustrations associated with face-to-face contact and thereby enables the meeting's ultimate purpose to be achieved.

The following are some key features of computer teleconferencing:

The computer provides sophisticated software support.

Other teleconferencing systems require real-time, geographic-dependent interaction, but CT can be used at any time from almost anywhere. Accordingly, meetings can be prearranged or scheduled like other teleconferences.

It is the least expensive of the teleconferencing technologies.

It can use creative software systems to develop models for better decision making. With the growing emphasis on decision support systems and management tools for graphic display of information, it can accommodate new modeling systems without radically changing user interaction.

CT provides the key advantage of long-term record and electronic filing. Large volumes of file records can be accessed, like an electronic library, and can be easily and rapidly retrieved.

No special computer terminal training is required with online "help" and simple commands.

Communication is improved among managers. The network becomes a "place" in the thought processes of those who are connected via computer communications.

Turnaround time is reduced on urgent decisions or actions. (In one U.S. Army test using computer conferencing, decisions that would have taken one to two weeks took one to two days in many cases.)

Interruptions from telephone calls are reduced because information can be sent whenever the recipient is free.

People are never late for a meeting. They use the system whenever and wherever they find it convenient.

Messages and information are organized for the most logical presentation. The discipline of putting thoughts into writing before communicating them improves the quality of communications.

Manager stress is decreased because managers are always connected to the office, never out of touch.

System production of copies (instead of making multiple photocopies) is improved.

Techniques for solving problems which require action and review by several people and activities are more effective.

Electronic "footprints" can keep everyone involved in the project informed—from beginning to end. People can enter the process at any point and have full documentation to evaluate the process.

Training of new staff is accomplished at a lower cost per student. Current staff members do not lose time away from the job.

Management/employee relations are improved for those affected by a policy document because these workers can digest the material easier at its successive stages of development.

Scheduling problems are eliminated.

Computer conferences can also be synchronous in some systems, and this option influences other patterns of usage as mentioned. Face-to-face conferences are synchronous; everyone must be in the same place at the same time.

Groups can convene electronically, then decide to meet or break up into committees of various sizes. While the larger conference continues, each member can electronically take notes on computerized scratch pads and can send these notes to another member without disturbing the conference.

Medical diagnosis is easier. The doctor sends out a description of the problem to all subscribers of the system in order to receive assistance.

Far from discouraging personal participation, a computer conferencing system encourages users to write ideas as they occur. This gives users time for reflection, which might not be available in a face-to-face meeting. When computer teleconferencing is used to prepare for face-to-face meetings, the background of issues and trivia have been dealt with, and people are ready to make decisions or move forward on more difficult questions.

ELECTRONIC MAIL VERSUS COMPUTER TELECONFERENCING

In most electronic communication systems, there are four levels of involvement, access, or use:

Bulletin board—open or global access.

Conferencing—controlled or group access.

Electronic mail—one-to-one access.

Memo or scratch pad—private access.

See Figure 9–11.

Bulletin Board Features

Bulletin board access is generally available to all users. Although participation in conferences is limited to being asked or selected, electronic mail is useful for short messages, such as "Let's have lunch" or "Did you get that report finished?" The memo area becomes the electronic desk.

Electronic Mail Features

This tutorial on electronic mail represents some, not all, of the features found in electronic mail systems. Electronic mail systems do not all have the same organization, just as Fords and Chevrolets are automobiles but do not look at all alike. For example, electronic mail produced on word processors is usually transmitted in real time, whereas store-forward mail uses a central computer to deposit mail, and retrieval is accomplished in nonreal time. This technique may be a solution for those plagued with "telephone tag" or "telephone Ping-Pong," the seemingly interminable calling

FIGURE 9–11

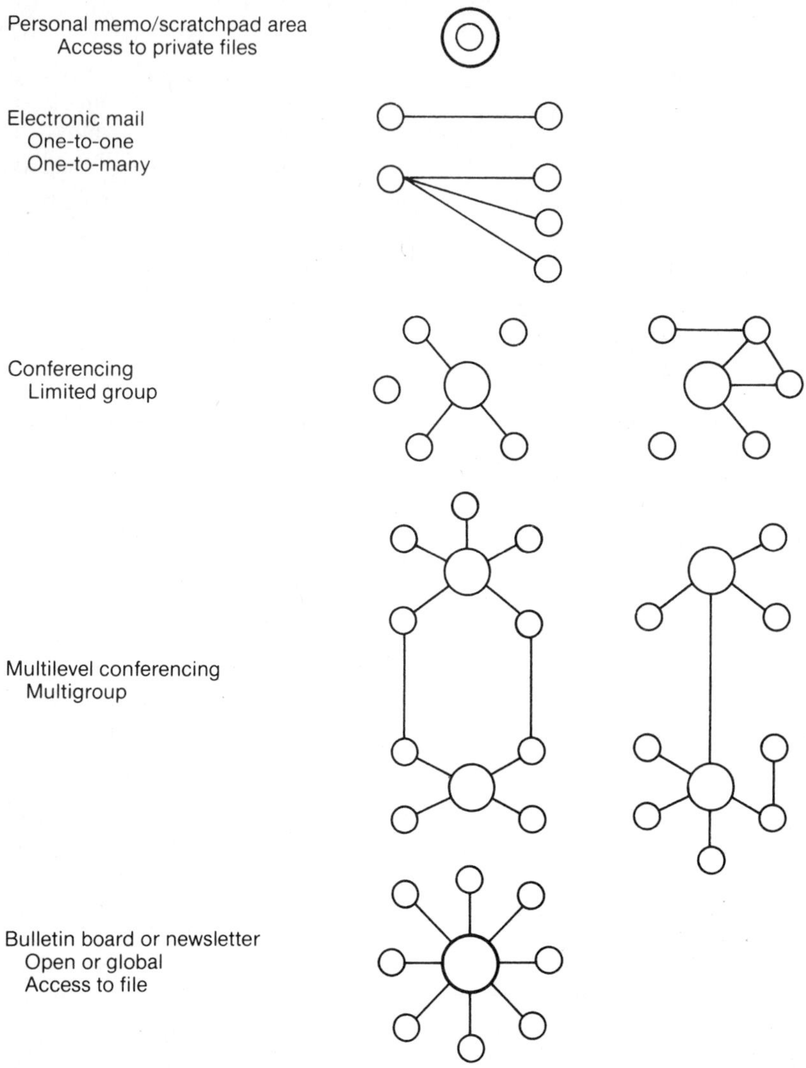

Personal memo/scratchpad area
  Access to private files

Electronic mail
  One-to-one
  One-to-many

Conferencing
  Limited group

Multilevel conferencing
  Multigroup

Bulletin board or newsletter
  Open or global
  Access to file

Courtesy: Cross Information Company

back and forth before one reaches another person. Moreover, ideas can be sent or broadcast to groups of people. Ideas can also be networked with other personal idea points for idea collection.

Unlike the U.S. Postal Service, there must first be a recognized recipient on the electronic mail system before a message can be sent. A formal game is required in order to create a message, and most mail systems also have certain protocols for creating a message—usually requiring participants to access certain commands, such as ".R" (period commands), which allow one to review a message before sending it.

The following discussion will focus on the store-forward approach. Some of the commonly found electronic mail commands for sending are:

> *Send message*—puts message in receiver's pigeonhole on the computer.
>
> *Forward message*—allows a message to be read and passed along to one or more readers.
>
> *Quit*—stops activity and allows the user to leave the mail system.
>
> *Help*—brings up information on using the features as well as online instructions.
>
> *Timed message delivery*—allows the writer to create a message for future delivery. This feature can be used as a timed reminder to the writer as well.
>
> *Group*—allows distribution to a group of people.
>
> *Copy*—sends carbon or blind copies.
>
> *Registered/forced reply*—allows the sender to be informed when a message has been received and in some cases forces the receiver to reply to the message.

These features differ slightly with each electronic mail system. Some systems provide the writer with a simplified menu listing commands, whereas others provide a complicated yet comprehensive listing of commands. The analogy is ordering an entree or ordering a la carte.

Some of the typical features for reading or receiving mail are:

> *Acknowledge received*—lets the sender know that the message was received and allows a reply.
>
> *Again*—allows the message to be read again.
>
> *Print*—tells the computer to print the message out on a hard-copy printer.
>
> *Hold*—indicates that the recipient received the message, but there is no reply at this time. It also allows the recipient to hold the message until later.
>
> *Save/file/copy/delete*—puts the message in a permanent filing area or in the "waste basket."
>
> *Append*—allows a note to be attached to the bottom of a previously written message.

Reply—invokes the text editor so one can create a reply to the sender.

Forward—routes the message to a designated person for consideration. A message can be appended with comments, then be sent on to its final destination.

A myriad of other features found on electronic mail systems provide for multilevel passwords, listings of messages sent or received, verification of messages, and priority position. Standards like the CCITT X.400 will allow different electronic mail systems to be connected together.

Computer Teleconferencing Features

This new technology is called teleconferencing to differentiate it from such traditional office automation functions as word processing and simple, one-to-one electronic mail. From the perspective of the creator or writer, computer or text teleconferencing allows documents to be shared more easily and information to be exchanged on a group basis just like face-to-face meetings. Meetings can go on as long as necessary.

Computer teleconferencing (CT) systems also provide the following other benefits:

■ *Outlining.* A computer is ideally suited for generating and organizing an outline of unlimited length. It has the ability to add, delete, modify, and reorganize. In addition, most paper-based outlines do not allow for the easy addition of information. CT software offers the ability to "link" separate outlines together at any point in a grid form, which can be displayed on the computer terminal. In this way, related ideas can have distinct primary relationships as well as tertiary connections to other outlines. The prioritize command organizes a list of topics by their importance. See Figure 9–12.

The randomize command, for example, puts a list of topics in random order for possible associations. Grouping concepts by common attributes permits general observations. The categorize command creates an outline structure from grouped topics on a list, and other commands provide additional links to organize, identify, flag, and retrieve information.

■ *Charting.* The problem with most chalk or "white board" is that you often want to save what has been written, but need the space for something new. CT systems are ideally suited for saving work, while allowing for the creation of new text. In addition, the networking capability allows text to be sent to other people for review, comment, and correction. A "white board" of text can be sent to other project members for their comments. See Figure 9–13.

FIGURE 9–12

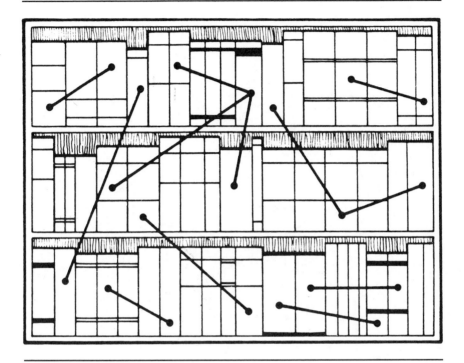

- *Networking.* Building personal "idea networks" is an important approach to creating and managing projects or management activities. The network allows ideas to be coordinated, managed, reorganized, and presented for review. Networking permits easy filing, searching, and editing. Creating ideas, linking them through multidimensional outlines, and charting their relative positions is certainly a far more exciting and creative process than the traditional filing system.
- *Idea exchanges.* An "idea exchange point" is an organized information outline for group networking. Each idea exchange can be divided into subcommittees, projects, chapters within a book, lectures within a course, or issues within a discussion, which are called idea points. Idea points let a group share information in a central idea exchange area or allow many-to-many or group networking. These areas are

FIGURE 9–13

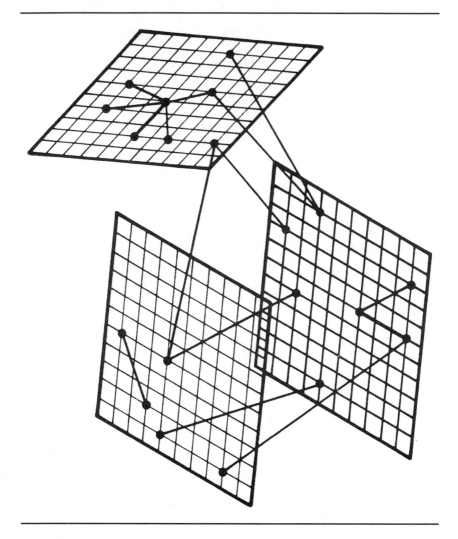

Source: Cross Information Company. Reprinted from James R. Weidlein and Thomas B. Cross, *Networking Personal Computers in Organizations* (Homewood, Ill.: Dow Jones-Irwin, 1986), © Dow Jones-Irwin, 1986.

accessed much like normal meetings, except there can be an unlimited number of idea exchanges (main topics) with a corresponding unlimited number of idea points (subtopics).

- *Status and tracking.* This function provides the conference member with information about new ideas that have been sent to the conference, new mail, and information that concerns and "keeps track of" project activities.

■ *Management reports and directories.* This feedback informs network members as to how long an individual has been working on an activity. This is useful information for those who work at distant locations, for charge-back systems, and for management reports. The directory is a useful system for discovering what idea exchanges and what members are on the system, or who may be interested in a specific issue.

■ *Searching.* Text searching of outlines, networks, conference members, discussions, and personal note pad areas can save large amounts of time as well as provide idea exchange participants far better organization. In addition, after a search is completed, the results can be instantaneously sent throughout the system. Searching, for example, generally allows a person to find comments on Venus and Mars but not Jupiter or Saturn. The system then retrieves the information requested and presents the conference member with material that can be organized into a report or filed. Joining merges text by lines, words, or sections. Dividing reduces text into lines, words, or sections. Sorting performs a variety of activities by arranging text in ascending or descending order, by crosspoints, charts, or other options.

■ *Gathering.* Once information has been amassed in separate idea folders, it can be reorganized into a summary report via the gathering function. This function collects personal memos, mail, and ideas, and organizes them in any way desired. Inserting and deleting allows input of new text within existing or new areas. Most CT systems have commands for text manipulation, such as move, copy, put, get, mark, and undo (which reinstates a command previously performed). Editing functions offer a wide range of options, including interfacing word processing software and internal editing commands. Some of the commonly found editing commands include insert, overtype, block move, and annotate modes.

■ *Real time.* Most CT activities are actually performed in a nonreal time environment, but this feature provides for an online conversation mode. The system "knows" who is presently "on" the system and can help organize a conversational idea session similar to a telephone conference or face-to-face meeting, except that statements are made in text form. In addition, a verbatim transcript of the event is automatically documented.

■ *File handling.* This is an increasingly demanding operation. Most systems can read ASCII files, but some have limitations on the amount of text transferable. Some systems allow for the uploading and downloading of ASCII files but cannot read them in the same form in which they were written. Printing options include spacing, selected files, outlines, charts, margins, headings, annotations, numbering, and status reports. See Figure 9–14.

■ *Other program options.* CT systems have to be user friendly and allow multiple modes of operation, menus for instruction, online help and tutorials, direct commands, multilevel string commands, access to the operating system, and user-defined macros. They must run either system programs or external timesharing programs, windows, and other administrative management features. Other features include online applications and simulations, both of which facilitate user interaction.

Features of Note or Scratch Pads

Each writer has private, secure (optional password protection) idea files or memo areas, which are kept online for ease of use. Essentially, idea files serve as an electronic desk having "folders" that contain memos, plans, and correspondence. This information can be sent to participating members and/or to the idea exchange.

Within the next five years, there will be more than 30 million personal computers installed worldwide, many connected via networks to corporate offices from either field or home offices. The WANG Personal Image Computer (PIC) combines video with the personal computer. Continued expansion of the personal computer market will bring changes in the teleconferencing industry, especially in small-scale desk teleconferencing and computer teleconferencing.

For effective problem-solving, managers should examine all teleconferencing technologies and understand how their capabilities may be integrated to enhance an organization's effectiveness. All teleconferencing forms facilitate effective communications over distances, and computer teleconferencing in particular allows communication to take place without regard to time.

MEETING ELECTRONICALLY, NOT ELECTRONIC MEETINGS

Computer teleconferencing, like a gatekeeper, recognizes certain user accounts as privileged and provides security and privacy for them. In an open teleconference, all members may read all items and contribute their own at will. Normally, they are not permitted to change files created by others, but this editorial function is up to the system designer. Many permutations are possible, and it is frequently useful to give at least one member editorial power to add and delete text.

A single computer teleconferencing system may contain numerous ongoing teleconferences. Subconferences can be open to anyone with a password to the system or restricted to a specific group. In either case, the software will keep track of who belongs to which conference and within each conference

FIGURE 9–14
The Electronic Communications Environment

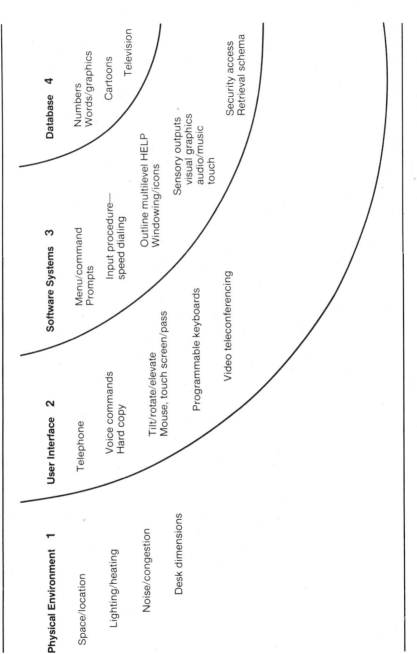

Physical Environment 1

Space/location

Lighting/heating

Noise/congestion

Desk dimensions

User Interface 2

Telephone

Voice commands
Hard copy

Tilt/rotate/elevate
Mouse, touch screen/pass

Programmable keyboards

Video teleconferencing

Software Systems 3

Menu/command
Prompts

Input procedure—
speed dialing

Outline multilevel HELP
Windowing/icons

Sensory outputs
visual graphics
audio/music
touch

Database 4

Numbers
Words/graphics

Cartoons

Television

Security access
Retrieval schema

Source: Reprinted from James R. Weidlein and Thomas B. Cross, *Networking Personal Computers in Organizations* (Homewood, Ill.: Dow Jones-Irwin, 1986), © Dow Jones-Irwin, 1986.

it will log who has seen which items. In addition, the system will notify the logged-in user when new material is present.

In computer teleconferences, the relation between subsystems is characterized by a strong semantic component. Although computer teleconferencing software looks like any other applications program, its structure reflects interaction patterns of explicit rather than implicit speech coventions. The foreboding task of learning to use computer teleconferencing is simplified by the presence of menus, maps to the logical network, and online help to guide the novice. Rather than a jumbled series of messages flying back and forth, computer teleconferencing is a highly organized, accessible form of communication. This well-instructed process incorporates the following components of computer software:

- Specified roles for teleconference participants that permit or restrain access to information, ability to vote on issues, etc.
- Selective communication among individual teleconferees.
- Capability to poll participants, with automatic analysis and feedback of results.
- Teleconferees can participate either under an assumed name or anonymously, thereby producing a "task oriented," rather than "status oriented," meeting.
- Capability to access, file, cross reference, and retrieve information from the computer.

Computer teleconferencing is more quick, more efficient, and more reliable than any other form of communication, including face-to-face. With its speed and efficiency, this tool will not only enhance each individual's productivity but will breathe new life into organizations that use it. (See Figure 9–15.)

FIGURE 9–15
Key Advantages of Computer
Teleconferencing

- No time restrictions—never late for a meeting.
- No geographical restrictions—always there.
- Low cost.
- Self-documenting and filing—electronic footprints.
- No actors or performing skills required.
- Self-pacing and training—online help and training.
- Convenient participation—on the road.

■   CONCLUSION   ■

In conclusion, below are listed the key advantages of teleconferencing:

Improved communications.

Improved time management and decreased management downtime.

Increased productivity.

Improved staff morale—allowing additional people to participate.

Increased speed of communications for more timely decision making.

Reduced delays due to meeting scheduling.

Lowered organizational and travel costs.

The benefits of teleconferencing to the corporation are obvious. However, a complex of ingrained habits (face-to-face meetings), old methods, technologies, and working environments could forestall quick implementation of teleconferencing. In nearly all cases, executives must forego some amount of travel, corporate "perks," time out of the office, and travel status.

Executives or users should understand that teleconferencing is not a direct substitute for corporate travel but a means to supplement communications, enhance productivity, improve decision making, and facilitate communication. Teleconferencing must be perceived as a means of improving the overall "quality of life" in the corporation, rather than as a way to eliminate enjoyable travel to other cities.

In the future, intelligent buildings and personal computers will play a vital role in teleconferencing. In addition, shared teleconferencing centers, provided to tenants of intelligent buildings, will streamline adoption of this new technological concept. See Figure 9–16. Teleconferencing has special applications in long-distance communications, nationally and internationally.

Teleconferencing is a new management, education, and communication tool. It has been proven extremely effective in thousands of situations worldwide where it is designed and properly supported. Teleconferencing stands today where data processing stood 25 years ago. Many knew then what data processing could do but did not understand how to do it. Teleconferencing has been growing steadily, but quietly, for the past five years. During the next five years, it will grow by quantum leaps, becoming a tool that will make companies more competitive, more effective, and more profitable.

IMPLEMENTING TELECONFERENCING

There are as many plans for teleconferencing as there are diet aids. What we need is a successful diet plan for the new electronic fashions of the 80s. Many present plans take the "cucumber diet" approach, requiring one to

FIGURE 9-16

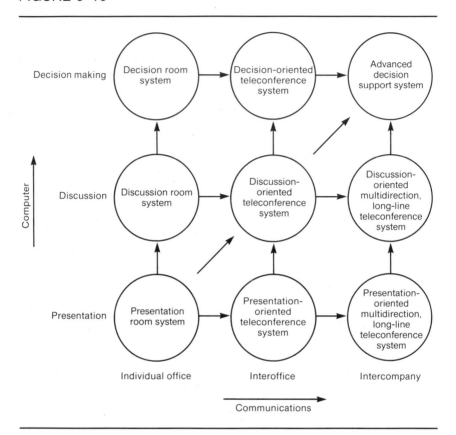

Courtesy: NEC

eat nothing but cucumbers, until one resembles a cucumber. In a similar way, teleconferencing users find themselves swamped with vendors purveying one theme: "Listen to our plan; the HAL 9,000 is the only system you will ever need; ours is the 'systems' company." Each of these tactics is the same old snake oil approach to solving problems, problems you didn't even know you had!

There remains another problem; some people hope that interest in teleconferencing will go away. Then they can get back on the plane and go home.

Implementing any new technology requires an enormous amount of time and energy. The following steps are key suggestions to implementing a teleconferencing system.

1. *Start now.* Too many organizations play around with the decision of who should captain the ship and thus never seem to get out of port. If plans are kept simple, and pilot projects move along, progress will be made, albeit slowly. Most important, take teleconferencing in small bites. Examine one operation, figure out if it works, modify it, learn about it, expand it, and continue to move forward. Someone needs to pilot the project. Without this effort the ship will wander in circles and most likely sink.

2. *Think small.* Do everything in small steps with incremental additions. Find pockets of people interested in getting started, work with them for a while, then find another group, using the first as a model.

3. *Let the monster sleep.* If you find people who don't want teleconferencing, ignore them. Teaching them is like trying to teach a pig to sing— it wastes your time and annoys the pig. These people may change, but more likely technology will change to be more compatible with them. See Figure 9–17. Technology, in fact, changes much faster than we think. People, on the other hand, change much slower.

4. *Understand the enemy.* The difficulty in adapting a new technology resides in the job performed. If worker titles are examined, it is plain to see that nearly everyone has a different job title and specialized information that goes with it. The simple reason robotics and factory automation have been successfully incorporated into the manufacturing environment is that by programming machinery to assume repetitious tasks, workers are freed to pursue more demanding undertakings. As we move into higher and higher management positions, jobs become vastly more complex and more dependent on an individual style of management. Here, automation can hope to be an "executive helper" at best. Teleconferencing is one of these helpers. It is a supplement to travel rather than a replacement. It should be viewed as a management "power tool" to help managers gain visibility, manage workloads, increase work quality, and so forth.

5. *Sell the sizzle.* Marketing is the key to office automation. *Organizational marketing,* a term coined years ago, describes what is much needed today: The tactics and strategies used to sell products and services within an organization differ little from those used by vendors. Motivation techniques, posters, ads, promotional devices, and (even) cash incentives are needed to encourage in-house utilization as much as outside marketing.

6. *Train the beast.* Training will be a limiting factor until vendors develop electronic learning systems. Training people electronically is faster and suits personal time schedules better. According to noted writer Clifford Barney, "One benefit of electronic training is it gets you hooked up to the keyboard so you learn about the terminal and realize it won't

FIGURE 9–17

Courtesy: Cross Information Company

bite." Teleconferencing suffers from an enormous failure rate. It can be compared with a drivers education course all of whose students kill themselves in auto crashes! People think teleconferencing is going to be great, but when they fail, they give up. People should be taught how to fail, or at least how to recognize an accident. Then they can learn how to make teleconferences safer.

The biggest problem with teleconferencing appears to be user attitudes. For example, lack of body language, which provided cues in traditional meetings, is a major stumbling block to adoption. The telephone does not make use of body language, but the human voice communicates emotions and attitudes.

Some of the problems associated with most forms of teleconferencing include:

- Unwillingness to learn new procedures.
- Fear of proper equipment operation and connection.
- Fear of failure when conducting a teleconference.
- Unwillingness to organize material.

To proceed with implementing teleconferencing, it is necessary to develop a "business case."

The Business Case

In nearly all teleconference installations, there is a commitment by a senior executive prior to implementation. This is because of the typical high-capital cost, which requires top-level approval. However, management does not commit corporate resources without a corresponding requirement to use the equipment.

To implement teleconferencing these steps are necessary:

1. Establish corporate objectives.
   - Lay the groundwork with upper management.
   - Build the foundation with end users.
2. Determine real versus perceived benefits: Increased communications, shorter decision times, increased productivity, lowered travel and management downtime.
3. Implement.
   - Check vendor lead times.
   - Check internal facility availability.
4. Analyze user requirements.
   - User needs are primary.
   - Analysis is essential to meet those needs.
   - Use aids in finding potential users.
   - Determine real need.
   - Decide who will do analysis—in-house or consultant.
5. Do a communications audit. A user audit would include analysis of the frequency and nature of the following:
   - Modes of communication—voice, data, video/graphics, and text.
   - The communication network—internal office, intrabuilding complex; regional sites, subsidiaries, branches; external, customers, audience, or membership; other business and professional connections.
   - The information network—access to information, storage of information, transmission of information, information-float waiting times.

- Technology issues—present office tools, yellow pads, calculators; smart office tools, personal computers, teleterminals, and so on; office support systems, electronic mail, voice storage, micrographics, office supplies.

6. Establish the life-support system.
   - Comfort—heating, cooling, lighting, humidity.
   - Ergonomics—sitting, standing, moving about.
   - Psychological—visual factors, convenience, flexibility.
   - Personal—parking, exercise, day care, windows, dining, social.

7. Establish the work and meeting matrix.
   - Time required to plan.
   - Types, numbers, and relative importance.
   - Need for and use of graphics.
   - Nature of communication—real versus nonreal time.
   - Desire for flexibility in communication.

8. Decide on the travel matrix.
   - Frequency.
   - Destinations, distances.
   - Exact departure and return times.
   - Attitudes toward travel—telecommuting options.
   - Opportunity, costs of working with people at a distance.

9. Analyze costs.
   - Telephone expenses.
   - Mailing costs.
   - Travel expenses, including actual (not only budgeted) expenditures for automobile expenses (both local and remote); aircraft rentals, fares, baggage charges; entertainment, tips; hotel room, meals; public transportation, parking; telephone tolls; interest lost on travel advances and on prepaid deposits; unexplained expenses; cost of time spent on travel.
   - Review of all the other costs from life-support systems to software systems for the worker.

10. Analyze the prospects.
    - Determine potential user groups.
    - Develop methodology based on those group activities.
    - Topics for investigation are: travel information, meeting information, familiarity with teleconferencing, interest.
    - Learn how they do business.
    - Understand their needs.
    - Describe benefits.
    - Suggest uses tailored to them.
    - Query their perceived needs.
    - Don't expect complete understanding.
    - Expect resistance.

- Have patience.
- Conduct extensive interviews.
- Analyze results.

11. Evaluate the economics.
    - Prepare introduction.
    - Outline business case issues.
    - Provide financial analysis summary: capital expenses, implementation costs, recurring costs, displaced costs, comparison of alternative cases, sensitivity analysis.
    - Present system considerations: obsolescence, compatibility, room design, service and support costs, training required by teleconference staff, backup and alternatives, system security.
    - Develop and present audit trail criteria.
    - Develop management and organization impacts.
    - Present teleconference team and responsibilities.

12. Introduce the system.
    - Begin the real work: Early awareness should take place now; bring in more help, if needed.
    - Promote. Name the system—like a boat, for notoriety. Handle it as a new system. Advertise in company newsletters, local magazines, and posters on bulletin boards. Hold a "grand opening," complete with monthly receptions for new staff. Talk it up! Call those people originally interviewed and get them involved. Get help from within the company through corporate communications, graphics, marketing. Include a telephone number for information. Develop series of user manuals, such as a pocket guide, a technical manual, and a sales brochure.
    - Training. This essential part of the system must be geared to users and must hold their attention. Speak to the training department for help. Keep training short, sweet, and graphic. Use videotape.
    - Followup helps to ensure repeat usage. It takes four or five teleconferences to understand benefits, and the real measure of success is repeat usage. Observe teleconferences if possible, and hold users' hands and guide them. Keep it up until they are convinced because satisfied users are the best advertising. Ask for candid reactions.

13. Make expansion plans.
    - Modify the room or system according to users' suggestions.
    - Look for new users, applications, high-travel areas.
    - Continue advertising.
    - Write up progress for internal company organs and trade press.
    - Monitor usage, comments, trends, and technology.

The business case is an important planning tool. It gives you a benchmark from which to develop new systems or to expand existing ones.

■ REFERENCE MANUAL: RESOURCES   ■

Industry Association
International Teleconferencing Association—IT/CA
  1299 Woodside Drive, Suite 101
  McLean, Virginia 22102

Consulting and Training
Intelligent Buildings Corporation
  1881 9th Street, Suite 311
  Boulder, Colorado 80302
Virginia A. Ostendorf, Inc.
  P.O. Box 2896
  Littleton, Colorado 80161

Computer Teleconferencing
Advertel Communication Systems
  2067 Ascot
  Ann Arbor, Michigan 48103
Cross Information Company
  1881 9th Street, Suite 311
  Boulder, Colorado 80302
Infomedia
  801 Traeger Avenue, Suite 275
  San Bruno, California 94066

Audio Teleconferencing
CONNEX International
  12 West Street
  Danbury, Connecticut 06810

Confertech
  8795 Ralston Road
  Arvada, Colorado 80002
The Darome Connection
  5725 East River Road
  Chicago, Illinois 60631
WESTELL
  7630 Quincy Street
  Willowbrook, Illinois 60521

International Teleconferencing
INTELSAT
  490 L'Enfant Plaza, S.W.
  Washington, D.C. 20024
COMSAT General
  950 L'Enfant Plaza, S.W.
  Washington, D.C. 20024

Narrowband Video Teleconferencing
Colorado Video
  Box 928
  Boulder, Colorado 80306
NEC America, Inc.
  2740 Prosperity Avenue
  Fairfax, Virginia 22031
Robot Research
  7591 Convoy Court
  San Diego, California 92111
GEC—Jerrold
  2200 Byberry Road
  Hatboro, Pennsylvania 19040

Full-Motion Video Teleconferencing
Private Satellite Network
  215 Lexington Avenue
  New York, New York 10016
Satellite Business Systems
  8283 Greensboro Drive
  McLean, Virginia 22102
AT&T Communications
  Route 202/206
  Bedminster, New Jersey 07921
Centro Corporation
  9516 Chesapeake Drive
  San Diego, California 92123
American Satellite Corporation
  1801 Research Boulevard
  Rockville, Maryland 20850
Pierce Phelps
  2000 North 59th
  Philadelphia, Pennsylvania 19131
Environmental Systems Integration
  Corporation
  7300 York Road
  Baltimore, Maryland 21204

Graphics and Telewriters
Optel Communications
  90 John Street
  New York, New York 10038
Interand Corporation
  666 North Lake Shore Drive
  Chicago, Illinois 60611

Limited-Motion Video Teleconferencing
Compression Labs
  2305 Bering Drive
  San Jose, California 95131
Widcom
  1500 Hamilton Avenue
  Campbell, California 95008

Teleconferencing Newsletters
TELESPAN
  50 West Palm Street
  Altadena, California 91001
TELCOMS
  Old Radio Hall
  975 Observatory Drive
  Madison, Wisconsin 53706
TELECONFERENCE
  Applied Business Communications
  5 Crow Canyon Court, Suite 209
  San Ramon, California 94583

Other Teleconferencing Services
AT&T Communications
  AT&T Room 3A120
  Bedminster, New Jersey 07921
AT&T Information Systems
  P.O. Box 6835
  Bridgewater, New Jersey 08807
Bonneville Satellite
  165 Social Hall
  Salt Lake City, Utah 84111
Cable and Wireless, Ltd.
  Teleconferencing Manager
  420 Lexington Avenue, Suite 2020
  New York, New York 10170
Communications Training Consultants
  450 Park Avenue
  New York, New York 10022
Comtech Data Corporation
  350 Hayden Road
  Scottsdale, Arizona 85257
CENTRO
  9516 Chesapeake Drive
  San Diego, California 92123
Communications Training
  450 Park Avenue
  New York, New York 10022

Confernet
  10960 Wilshire Boulevard
  Los Angeles, California 90024
Digigraphic Systems
  10273 Yellow Circle Drive
  Minnetonka, Minnesota 55343
Digivision
  11722 Sorrento Valley Road
  San Diego, California 92121
GTE International Systems Corporation
  140 First Avenue
  Waltham, Massachusetts 02254
Honeywell Video Graphics
  121 Presidential Boulevard
  Bala Cynwyd, Pennsylvania 19004
Hughes Television Network
  Teleconferencing Manager
  4 Penn Plaza
  New York, New York 10001
ITT World Communications
  100 Plaza Drive
  Seacaucus, New Jersey 07096
HI-NET Communications
  3742 Lamar Avenue
  Memphis, Tennessee 38195
Hilton Hotels
  9880 Wilshire Boulevard
  Beverly Hills, California 90210
MARCOM
  P.O. Box 66507
  Scotts Valley, California 95066
MONTECH
  2110 East Roxboro Road, N.E.
  Atlanta, Georgia 30324
NEC America
  2740 Prosperity Avenue
  Fairfax, Virginia 22031
Northern Telecom
  111311 Winners Circle Drive
  Los Alamitos, California 90720
Pacific West Communications
  203 Auto Drive South
  Compton, California 90220

Public Broadcasting Service
Teleconferencing Manager
475 L'Enfant Plaza, SW
Washington, D.C. 20036

Public Service Satellite Consortium
1660 L Street, N.W.
Washington, D.C. 20036

RCA Global Communications
60 Broad Street
New York, New York 10004

Rapicom
3001 Orchard Parkway
San Jose, California 94134

Satellease
666 Dundee Road, Suite 1304
Northbrook, Illinois 60062

Satellite Networking Associations
10 East 40th Street, 23rd Floor
New York, New York 10016

Shure Brothers
222 Hartrey Avenue
Evanston, Illinois 60204

TeleConcepts
145 East 49th Street
New York, New York 10017

TRT Telecommunications
1747 Pennsylvania Avenue, N.W.
Washington, D.C. 20006

Videostar
3390 Peachtree Road
Atlanta, Georgia 30326

Video Systems
P.O. Box 12901
Overland Park, Kansas 66212

Vidicom
742–D Hampshire Road
Westlake Village, California 91361

VITALINK
1330 Charleston Road
Mountain View, California 94043

Wold Communications
10880 Wilshire Boulevard, Suite 2204
Los Angeles, California 90024

■   RECOMMENDED READING   ■

Teleconferencing: Linking People together Electronically, by Kathleen Kelleher and Thomas B. Cross (Prentice-Hall: Englewood Cliffs, N.J., 1985).

Author

Thomas B. Cross
Managing Director—Cross Information Company
Vice President—Intelligent Buildings Corporation
Boulder, Colorado

Thomas B. Cross is one of the foremost authorities on the technology, implementation, and management of teleconferencing, telecommunications systems, and intelligent buildings. His company, Cross Information Company, developed MATRIX, a computer teleconferencing system that received the prestigious Outstanding Information Technology Award from the Associated Information Managers of Washington, D.C. He received the Distinguished Author Award from the International Facility Management Association. He was a founder and is a member of the International Teleconferencing Association (IT/CA) and Editor of the *IT/CA News*. Intelligent Buildings Corporation provides management and development of intelligent buildings in office parks.

Cross Information Company sponsors the following annual conferences: Intelligent Buildings Conference, International Teleconferencing Symposium, Telecommuting Technology Conference.

CIC is a research and development company involved in teleconferencing, telecommunications, and intelligent buildings. CIC provides management reports and custom product and strategic planning research, management seminars, and consulting. CIC develops advanced software systems, including CROSS/POINT™, which provides charting, outlining, electronic mail, conferencing, and other advanced networking features.

He has lectured, written, and consulted extensively on teleconferencing, intelligent buildings, and telecommunications in the United States and Europe. He has written more than 50 articles on teleconferencing and information technology.

Cross has authored *Teleconferencing: Linking People Together,* for Prentice-Hall; *Communications Networking,* a book on electronic mail for Scott, Foresman; *Teleconferencing Report* for Pergamon Infotech; *The Softside of Software,* on computer software for John Wiley & Sons; *Telecommunications Management—An Executive Perspective, Intelligent Buildings: Strategies for Technology & Architecture, Networking Personal Computers—An Executive Perspective,* and *Telecommuting: Work Strategies for the Information Organization* for Dow Jones-Irwin.

Chapter 10

Planning the Teleport

John Paul Rossie
Western Tele-Communications, Inc.

Outline

INTRODUCTION

The 1980s seem to be situated in the middle of a technological revolution that started in the 70s and will apparently extend for at least several more years to come. The technological changes, occurring with ever-increasing rapidity, make these times especially volatile for the communications field. But in the midst of the changes and advancements, certain trends and long-term manifestations seem to have been established that you of the future will have a much clearer perspective on. One of those trends (or is it a long-term manifestation?) is the teleport.

Teleports became an industry buzzword in the early 1980s, generally thought to be due to some urban planning studies generated for the city and state of New York by the New York Port Authority and the National Research Council. In the original study, which attempted to paint a picture and solve the problems of life in high-density population centers of the not-so-distant future, the teleport was seen as the central location for all communication activities, the information hub for the urban center. That concept was followed up on by the New York Port Authority and blossomed into what is now the New York/Staten Island Teleport, or more appropriately, The Teleport. So in these few short years since the early 80s, the teleport has evolved from a concept and buzzword to a reality that is still unfolding and defining itself in 1986.

The definition of the word *teleport* varies from place to place and from person to person, depending on the developer and the user group. However, to talk meaningfully in terms of teleport planning, we must here make an attempt to put some firm limitation and definition on the word. It is ultimately the *uses* of a communication complex that determine its definition—whether that communication complex happens to be called a teleport, an antenna farm, or a shared transmission facility—and it is through precise definition that proper planning can be accomplished.

By the mid-80s, several associations had been formed to bring together the various groups of individuals and users who share common needs and interests related to teleports. Those associations tend to define *teleport* as best suits the common needs of each group. *Teleport* is defined as a function of the various uses it is seen to fulfill now and in the foreseeable future. The definition then becomes the basis upon which the planning efforts rest.

Knowing how many teleports are currently planned or even operating in the United States or in the world is simply impossible because of the ambiguity of the definition itself. I know of several communication facilities that could easily be considered teleports under a variety of definitions, yet none of them is mentioned in the "exhaustive" listings put together by various individuals or associations purporting to compile this type of information. The reasons for this are numerous and are compounded due to the ambiguity of the very definition of the term.

When The Teleport began to attract the attention of the communication world, there were already several satellite communication installations across the country that, however loosely defined, might have been called teleports. As the word became popular between 1982 and 1984, many people realized that, given a proper definition, what they had been operating for several years were indeed teleports. They had often been seen as antenna farms, where shared-use arrangements prevailed. But the potential of a teleport, at least at one end of a spectrum of possible definition, is much more than merely an antenna farm.

What we hope to do in this chapter is to examine the full spectrum of what could be considered a teleport—from the small and mid-sized facilities operating a few satellite antenna systems for multiple users, to the massive real estate ventures that clearly intend to serve a far larger market. It may be interesting to note how the "real definition" of a teleport will change and grow over time, encompassing more and more variations, rather than restricting itself; but such is for the future to uncover.

THE PLANNING FUNCTION

Planning is the effort to design facilities for future use by employing available, contemporary tools and ideas that are necessarily and by definition limited in foresight. In short, the planning function is good guesswork along with good engineering. Planning requires a vivid imagination as well as a good feel for the pulse of the times and the direction of the industry, which enables one to anticipate changes in technology as well as changes in attitudes toward technology. Both will affect the ultimate direction and use of technological development.

In planning for a teleport, a myriad of considerations tend to flow, one from the other, all seemingly equally important. In general terms, a teleport is some sort of communications complex utilizing satellite antennas, but the range and scope of the services provided by that teleport, which must be planned for and integrated into a total system, will undoubtedly include a multitude of services. Ultimately, there will be services for voice, data, video, audio, and combinations of these, employing encryption, protocol conversions, signal processing, production, distribution, and so on. Here is where we find the necessity for a firm definition of the teleport, to set the limits of the desired services so that the planning effort can move forward to make certain that specific types of services are available when needed.

Definitions

A teleport can be many things to many people. To some, it is an office park with ultramodern communication facilities; to others it is a geographically common aggregated node of many telecommunications networks that employ satellites as the primary means of signal distribution. The definition of a teleport that seems to offer a broad interpretation and contains most of the necessary basic elements that are typical of all teleports is given below. After examining this definition, we will look at the broad range of applications such a definition can encounter, and we will observe the vast differentials that can be considered as incorporated services of a teleport. The following seems to be the best all-round, usable, and ideal description of a teleport:

> (tel' e port), noun: a site offering a protected radio frequency (RF) environment for the colocation of satellite communication antennas, which takes advantage of user aggregation through the sharing of common communication plant and real estate facilities. Such common plant items involved in the shared-use concept include buildings, towers, AC power, DC power, signal interconnect and distribution systems, and other commonly used general site developments, such as access roads, security, fencing, lighting, wiring, telephone, and other common utility services.

This definition does not attempt to fill in much detail about the kind and number of potential communication services that will eventually locate themselves in the teleport, and for that reason it is a broader and more tolerable definition. It describes only the location and suggests the manner of communication system operation.

Contrast and compare the above definition with that of the American Teleport Association (ATA), as stated in their literature of mid-1985:

> *A Definition of "Teleport" as Defined by the ATA:*
>
> "Teleport" shall mean or shall apply to any facility which uses satellites and associated facilities for the transmission and reception of telecommunications services and which meets the following minimum criteria:
> 1. Multiple satellite network (carrier) access.
> 2. Multiple fixed and licensed satellite earth stations with both transmit and receive capability (e.g. colocation of multiple earth station operations with both uplink and downlink capability).
> 3. Multiple service capability (meaning two or more of audio, data, facsimile, video, voice, etc. telecommunications services).
> 4. Multiple regional interconnect capability to terrestrial networks.

Here the definition is somewhat more restrictive, tailored to the special interests of a defined group of users, specifying "multiple service capability" and "multiple regional interconnect capability" and requiring the ability to "both transmit and receive." If a teleport were developed strictly for video services, it would not fall within this definition. Nor would the hypothetical teleport with 100 antennas that did nothing more than receive information from the satellites and distribute that information by other means. Because of those specifications, the definition is more limited and restrictive but specifically targets the needs of the select group of users in the ATA.

In planning and defining a teleport facility, one must remember that the development of the services will be user-driven in almost all cases. Simply stated, this indicates that the facilities that are put in place are those that are immediately called for by contracts—facilities that will be put to immediate use and revenue production. The only instance where this may not be the paramount rule is with those facilities so large that the level of investment capital allows for the immediate emplacement of several types of facilities based on pure speculation, where the ultimate utilization rests on the efforts of future marketing. Such facilities are literally built with speculation capital. Because of the intense capital resources that this type of venture requires, it will not be the general rule in teleport development. Nevertheless, much of the infrastructure needed to allow future user-driven growth to occur will be preinvested as common plant items.

As the planning effort moves forward, it is essential to keep as much potential in mind as possible, building not just for the immediate requirements, but projecting future growth and services. For instance, there is no reason to limit oneself when making application for zoning permits by requesting construction parameters for 9-meter antennas if that means that new applications must be made when that potential 13-meter antenna becomes a reality. Obviously, the initial zoning request should mention the height and placement of the pending 13-meter facility. Otherwise, a second round of zoning hearings may be required because the structure height of the 13-meter dish was not authorized at first go-around or because the mention of multiple facilities was not initially made. Many local agencies will require projections of 5 and 10 years, hoping to avoid this very problem. This is simply a matter of thinking through the details and making the planning effort a complete one.

Teleport Categories

At one end of the spectrum of teleports are those facilities where location is the primary factor and selling point. These teleports could be sitting out in the middle of the prairie or hidden in some mountain valley, wherever best meets the criteria of low radio frequency interference (RFI). Once located, such a teleport could concentrate on communication services only, offering spots to other users for the placement of satellite antennas. No buildings other than equipment shelters and maintenance shacks would be planned, since the design of the teleport would include some sort of backhaul system for use as a communications interconnect into the nearest population center, eliminating the need for many on-site personnel. A teleport such as this will generally be a small or mid-size operation, where the owners have little or no interest in real estate development or business office activities and services. Such bare bones teleports are planned to be communications facilities only and will generally grow slowly, by client demand. Space for the location of antennas might be leased to a client for the placement of customer-owned and operated equipment, or the Teleport owners might provide all services necessary, including planning, equipment, construction, operations, and maintenance. In either case, very little speculation investment is made.

At the opposite end of the spectrum is the office park concept, such as The Teleport of New York, where more interest seems to be on the real estate development and ancillary services and advantages and where communications plays a secondary role, used perhaps to attract customers and to enhance the potential of being a client in the complex. Some examples of ancillary services and advantages of this type of teleport can be as widespread as concern for the surrounding economic community, local job market opportunities, and local business growth of support services. As suggested earlier, these types of teleports are generally large projects with massive financial backing. The Teleport of Staten Island is a joint venture of Merrill Lynch, Western Union, and the New York Port Authority, a partnership that commands impressive monetary stature. A client at such a teleport might have communication services only as a secondary requirement and may be more interested in the office space and the proximity to the other businesses that will locate in the office park.

Because of the large communications traffic densities that are expected to be generated by such facilities (as in New York) and because of the huge capacity potential of fiber optic systems, the larger teleports often include fiber optic interconnectivity to major markets. Also, as we shall see, location near the large urban areas often precludes the use of interconnect microwave systems due to existing radio frequency congestion in the immediate vicinity.

Whatever the size or the main thrust of business of the teleport, the requirement for communication interconnection to the nearby urban service market is an essential element and must be an integral part of the planning process. Otherwise, the facility will strand itself through a communication bottleneck and eventually cut itself off from the rest of the world—an ironic fate for a facility constructed to enhance communications.

Purposes of the Teleport

In planning for teleports, one encounters what can be perceived as a new pattern of thinking that has emerged since the late 70s. This new attitude permeates many disciplines, not only communications; it can be characterized by such attributes as attention to environmental concerns, attitudes of cooperative effort between competing or even nonrelated industries in both the public and private sectors, and a search for innovative solutions to the many new problems brought about by high technology. Along this same line, the well-planned teleport offers a unique solution to common but serious problems facing the world of modern communications.

The primary concerns that should constantly lie behind the planning of a teleport are fairly numerous. The various solutions to these concerns offer several advantages, each aspect and solution of which can become a defining element of the concept of a teleport. For instance, the idea of shared use is a primary motivater in teleport development because of the obvious sharing of resources that can be brought about by such an endeavor. Even so, the communication facilities might still support the single user, such as a common carrier, since the nature of common carrier business does not always lend itself to multiple types of services from a single source. It might be that the carrier has chosen to locate in the teleport because of the convenience of other ancillary services. On the other hand, shared communications facilities will be popular in many instances, since multiple users can band together to run several services through a single set of equipment at a reduced cost.

The concept of a teleport also involves the idea of a centralized location sharing land-use permits, which solves the problem of multiple zoning applications for identical purposes. When an authorized planned usage development becomes an active project for the construction of a teleport, much of the work with local authorities is consolidated, making things easier for the developers, the planners, and the local zoning authorities.

Once the teleport project is underway, the shared plant is another attractive feature. Examples of such shared plant items include the building, air conditioning, AC/DC power facilities, interior and exterior lighting, fencing, security services, radio towers and antennas, uninterruptible power supplies (UPS) units for power protection, local distribution networks, general ground work and landscaping, and a thousand other things, both necessary and convenient, that will make teleport use easier and cheaper for everyone.

Another less obvious but equally important shared resource at a teleport will be the personnel, including design, engineering, marketing, operations, and maintenance personnel whose talents can be given a broader range of application by bringing more potential work to them at a single location. These people can be considered as conserved resources in the same way that we can consider the teleport as helping to conserve the radio spectrum through concentration of and aggregation of facility usage. Sharing in the expense of such valuable and costly resources means less individual consumption and effort. The bottom line, of course, is less capital expenditure per single user for many of these shared items, making the entire operation less expensive for all users involved.

The well-planned teleport will accomplish all of this while locating itself conveniently near an urban center to serve the needs of users who tend to aggregate in the nearby population area. Such convenience will itself inspire growth and fan the fires of demand, making the teleport a self-perpetuating kind of service center. However, teleport planning is not without its inherent set of problems.

COMMON PROBLEMS IN PLANNING

Radio Frequency Interference (RFI)

Common carrier microwave radio systems generally utilize the 4 GHz, 6 GHz, and 11 GHz radio frequency bands. C-Band satellite systems, by far the most popular type in use in the mid-80s, utilize 6 GHz for ground-to-space and 4 GHz for space-to-ground transmissions. Ku-band satellites, especially in the international services, utilize an 11 GHz downlink (space-to-ground). All of these systems, then, are interference potentials to other services, and efforts must be made to keep interference to a minimum. Good frequency coordination planning and ongoing frequency protection is essential.

Terrestrial microwave radio routes create a virtual lattice work of radio paths across the United States that act as interference sources and create areas and zones where transmitting earth stations are forbidden to operate due to their potential to cause interference to a nearby terrestrial system. (The converse is true of the earth station if it is licensed and in place first. In that case, the terrestrial system will be disallowed.)

In many cases, either the planned satellite earth station must then be located away from the interference zone (which unfortunately is usually near to the prime urban locations) or else some sort of shielding must be devised to block the radio energy transmitted into and out of the earth station site. A teleport, because of its multiple-user and shared-cost concepts, might more easily afford to construct the type of shielding that would be necessary to provide RFI blockage than could a single antenna user. Natural terrain blockage is by far the cheapest type. However, expense has not stopped the stout-hearted. The New York Teleport has recently completed the construction of a 60-foot high wall, built largely of concrete and steel, around an 11-acre plot to shield the planned satellite antennas at that location.

In some instances, natural or existing RFI shielding is available, such as hills and other permanent features of the terrain. In other cases, buildings, whether belonging to the teleport or simply nearby structures, might afford some blockage. However, any earth station located near urban development areas faces the possibility of this "free blockage" concept working against it. New building construction will create new reflection patterns. These new patterns can arise with new buildings that are put up long after the earth station has been in operation, thereby bringing into being a new source of interference through reflection that could seriously disrupt the operation of the earth station from that time forward. Good planning for teleports then requires that the site be currently shielded from RFI and that it have very little potential of developing future RFI problems for whatever reason.

Space Availability

Business applications for satellite-transmitted voice and data circuits are increasing in developed urban high-rise and suburban office park environments. Along with this increase are new services available, such as VSAT (very small aperture terminal) installations. Regardless of this possible competition for the teleport, many users will still insist on the larger antennas to meet their needs. Many potential users situated in such areas find that there is simply no available ground space for the placement of a satellite antenna, particularly if the size is larger than 5 or 6 meters. Roof mounting is a severely limited possibility because of structural problems, space limitations, and wind-loading problems at rooftop levels—and of course there is greater exposure to radio frequency interference (RFI) at the high, clear levels of an office building roof.

Teleports, when properly planned, offer placement for many antennas both large and small. Since this generally means location away from the ideal spot at the user's premise, interconnect systems utilizing microwave, cable, optical fiber, or other communication links then must be used to relay the signal to and from the teleport and the user premises.

Local Building Requirements

Even when the space is available, local building codes may not permit the placement of large diameter (or even small diameter) satellite antennas on or near the property of the intended user. The enforcement of such local zoning codes has been increasing in communities around the country based on a number of often misunderstood factors, including public safety or even local flavor and design-continuity considerations. In the mid-80s there are anywhere from 500 to 750 communities in the United States that place some restriction or other upon the construction of satellite earth stations. These restrictions for the most part are intended to apply to TVRO (television receive-only) systems, but many such ordinances affect the larger commercial installations as well.

Most building commissions require that the submitted plans at least bear the stamp of a local or state licensed design engineer who has reviewed the drawings being submitted. Having this local or regional approval before the initial presentation of the plans is a good time-saver, although an added expense.

Urban Placement Problems

Because of the highly dynamic environment of urban development, it is good to mention again some of the potential planning woes caused by new building construction that create havoc in the prediction of RFI patterns. As modern architectural trends develop, there seems to be more and more structural freedom, like the creation of massive curved surfaces of glass and other smooth, highly reflective materials. Square buildings were hard enough on frequency coordination, where the building sides acted like huge passive reflectors, throwing off the radiation in a loosely predictable direction. However, these modern structures do not allow even that luxury of loose prediction.

In addition, changes in land development and usage patterns create the potential for zoning changes and shifting local attitudes that could adversely affect the ongoing operation of an existing teleport. The best planning is to avoid these types of situations entirely and to keep the fate of the teleport free from as many serendipitous elements as possible.

Environmental Issues

Local and national action groups have been known to be active in restricting the development of such radio frequency emitting facilities as satellite transmission stations. Usually, this occurs when the potential teleport is located too near a residential area. As environmental concerns grow, it is likely that many areas of the country will experience local outcry against the often misunderstood (and little known) "dangers" of microwave radiation. However, the confinement of many sources of such radiation into the single area of a teleport away from urban and residential activity is likely to be much more palatable to some environmental groups.

Aesthetics

Many communities and even business park developments are restricting placement of satellite antennas for no other reason than appearance. To some, the round, ungainly, mushroom shape of the satellite antenna is offensive. The clash between high technology and the subjective criterion of aesthetics simply represents another area wherein the planner must contend with opposition. In some instances, local pressures have forced owners of satellite antennas to paint their antennas certain earth-tone colors in order to make them less conspicuous and more appealing to the local communities.

DESIGN CONSIDERATIONS

The actual layout and design of the teleport will be determined by several factors, few of them related to questions of engineering and technology. For instance: What will the primary services of the teleport be? Will there be growth into new services, and if so, how soon? How much land is available for both the initial phase and for expansion? What are the immediate geographical and RFI environments?

It would seem that once these questions are answered, one can move ahead to the next stage of planning, the actual layout and physical placement of equipment shelters, cabling, power lines, waveguide runs, conduiting, water and sewer runs, foundation placement, internal building arrangements, and the like. Each antenna will require full satellite arc visibility; each antenna should be positioned to create the least amount of RF and visibility interference to every other antenna; and each equipment shelter and system head end should be positioned to create the minimum wave guide distance from the antenna.

With those criteria in mind, a number of design possibilities present themselves. Each teleport will probably be laid out a little differently, based on geographical location relative to the section of the satellite arc that is of interest to the users. One ingenious approach to teleport design was devised by a creative marketing engineer from Western Tele-Communications, Inc., (WTCI), Gary McCue. McCue reasoned that from the standpoint of interference blockage, the optimum shape for a teleport is triangular, oriented with one long side nearly parallel to the equator and one angle pointed to the pole, that angle pointing north if the teleport is in the U.S. domestic service. In this particular triangular arrangement, a number of interesting things occur:

> As the geographical location of the teleport changes, so does the shape of the triangular outline. The length of the sides will be determined by the relative direction to the extremes of the orbital arc section of interest. Such change will be directly proportional to the local azimuth and look angle requirements. For example, the further south and centered under the arc the teleport is, the more equilateral the sides and the "fatter" the triangle, because the antennas will generally all be looking at a high angle, almost directly overhead.

> As the teleport location moves north and to the east, the elevation angle to the arc lowers, and the orientation to the center of the arc tends westward. The resulting shape is one of a shorter eastern side and a much longer western side, which is drawn out to a more closed angle. Internally, the antennas will need to be placed further behind one another to allow clearance over the top, which lengthens the requirement for land along the western and southern corner.

> The triangular shape allows for controlled expansion of the facilities, with the initial antennas located "high up" in the northern corner. The first equipment shelters can be built along the edges at the northern

peak, accomplishing two objectives: (1) This layout presents the shortest possible waveguide run between the antenna and the shelter, and (2) the building can create some bit of RFI blockage.

As the Teleport expands, the antennas spread toward the open area of the southern end, and more equipment buildings are built down the sides to accommodate the new antenna installations. Since the sides point roughly to the extremes of the arc, the antenna orientation is such that the buildings along the edge always afford some bit of blockage from RF interference from the sides.

The triangular design offers the potential for the optimum level of coantenna clearance, minimum waveguide distances to the shelters, utilization of artificial interference blockage, and a clear pattern for new antenna placement and general site expansion.

The triangle is certainly not the only usable shape for a teleport and may be more academic than practical. Nonetheless, whatever shape the teleport happens to take, it should be based on well thought out plans that will allow for future growth and development.

■   CONCLUSION   ■

There is now a World Teleport Association, established to bring some semblance of order and structure to the development of teleports on a world-wide basis. This organization may offer an avenue for global interconnectivity and international cooperation in the field of communications but will also create new and unique problems to be solved. What are the various signaling protocols used around the world, and how can they be reconciled with one another? What sorts of standards exist for transmission, and which will be chosen as the operative standard? What political barriers and restrictions will need to be brought down before the world becomes the Global Village we've heard and thought so much about? Will worldwide networking lead to cooperative efforts in other communication-related and high tech ventures, such as space platforms and space stations? Once the world starts talking, really conversing with itself, starts breaking down the artificial barriers of politics, and starts sharing information and ideas, how will we ever get it to stop?

Author

John Paul Rossie
Western Tele-Communications, Inc.
Englewood, Colorado

John Paul Rossie, B.A., M.A., M.S., F.B.I.S., has been involved in communication system planning and feasibility study efforts since the final days of the NASA–CTS satellite experiments in 1978. His expertise covers a wide range of activity from system planning to market development in areas of satellite as well as terrestrial microwave signal distribution for voice, data, and video services. He is currently manager of marketing for Western Tele-Communications, Inc. (WTCI), with responsibility for new service development in those as well as other areas. Rossie has been instrumental in the development of the concept of teleport services within WTCI as well as industrywide. He holds degrees in telecommunications, philosophy, and business management and is the author of numerous articles on satellite communication topics.

Chapter 11

Technical Aspects of Applying Telecommunications and Related Technologies to High Tech Real Estate

Joseph L. Stern

Stern Telecommunications Corporation

Robert E. Weiblen

Integrated Systems Planning

T. J. Theodosios

Integrated Systems Planning

Outline

Source: Reprinted from Alan D. Sugarman, Andrew D. Lipman, and Robert F. Cushman, eds., *High Tech Real Estate* (Homewood, Ill.: Dow Jones-Irwin, 1985) © Dow Jones-Irwin, 1985.

This chapter has been written to familiarize real estate and other nontechnical professionals with telecommunications transmission facilities and technology.

The term "transmission facilities" covers a broad area ranging from internal cabling and wiring of a building to satellite communications. In this chapter, we will first discuss cabling and wiring. In the past, cabling and wiring a building for communications was a relatively easy task for the planner; the local telephone company provided not only the material, but also the technical expertise and expense for the engineering and installation of the cable facilities. In addition, the cable outlay for a building was designed to handle primarily voice telephone conversations with only a little data communication for the occasional user. Today, however, because of the Bell divestiture, the high cost of cable, personnel, and the rapid growth in integrated voice and data telecommunications technology, cabling a building becomes an important and far more complex issue, which must be dealt with by real estate planners and managers.

Related topics discussed in this chapter include sharing of communications services in a multi-tenant building, multiplexing, local area networks, direct termination systems, and the concept of "bypassing" the local telephone company communications facilities.

CABLE AND WIRING PLANS

Planning for the cabling of a high tech building involves careful consideration of present and future uses, the application of known technologies, and the application of conservative safety factors.

Building Entry Conduits

An average building can be assumed to contain 500,000 square feet of floor space with a mix of tenants having portions of floors and an occasional multifloor operation. One telephone will probably be installed for each 200 square feet of floor space; thus approximately 2,500 telephone instruments would be required to serve the building. This density, one telephone per 200 square feet, is appropriate for consideration in Manhattan, downtown Chicago, Center City Philadelphia, as well as Dallas and Houston office building locations.

Assuming that there will be 2,500 telephones installed in a building and giving consideration to the variety of communications concessionaire service offerings available, we can choose a split between the local telephone company-supplied services and services supplied by others. For the purpose of this discussion, the local telephone company will provide 1,000 lines, and the other services will provide 1,500 lines. Also assume that the 1,500 telephones served by the private service might very well go to a central PBX that serves another 1,500 lines in adjoining or nearby buildings. This approach provides for flexibility in design and allows us to anticipate the size of conduits that would be required.

The experience of telephone companies in various cities is the best resource for choosing conduit sizes. As an example, New York Telephone Company planners advise that a single 4-inch ID conduit can accommodate a 3,600-pair cable on a fairly straight run, although 2,700-pair cables are generally the practical upper-size limit due to conduit bends and similar factors. A 3000-line PBX in this high tech building would require 1,500 pairs (2-to-1 line-to-trunk ratio) for connection to the telephone company's central office and one 4-inch ID conduit for incoming trunk lines to the PBX. One additional 4-inch conduit would also be required for eventual connection to the adjoining building or buildings taking service from this PBX system. In addition, still another 4-inch conduit should come into the building to carry the 1,000 lines carried from individual telephones directly to the central office. This now calls for three 4-inch conduits entering the building at ground level or lower.

Additionally, planning for a high tech building should include sufficient additional capacity to provide for future fiber optics network cables, future interconnections to other buildings, and spare conduits accessing the building from different sides, to provide for the redundancy of cables to cover emergency conditions—such as disruptions due to broken water mains and other disasters.

Accordingly, for this 500,000-square-foot office building, the building entry conduits should look like this:

Building Entry Conduits for a 500,000-Square-Foot Office Tower

Function	Size (inches)	Number Required
Tower PBX	4	1
PBX Interconnection to other buildings	4	1
Local telco lines	4	1
Future fiber optics	4	1
Future building interconnections	4	1
Spares	4	2
Total		7

Sizes of Frame, PBX, and Maintenance Rooms

A review of PBX equipment offerings indicates that a 3,000-line PBX system requires approximately 900 square feet of floor space. To provide space for the frame room, power supplies, and maintenance facilities, an area of approximately 1,300 square feet should be reserved for the entire PBX system. The 1,300 square feet would be used as indicated in the table below:

PBX Equipment Floor Space for 500,000-Square-Foot Tower

Functional Group	Typical Room Size (feet)	Floor Space (square feet)
Frame room	15 × 15	250
PBX switch and associated electronics	25 × 25	650
Power supply, batteries, and air conditioning	17 × 17	300
Maintenance facilities, customer services, etc.	10 × 10	100
Total		1,300

Equipment floor loads vary substantially; the heaviest loads result from the standby power supply batteries in the PBX power supply room. Typical floor loads are tabulated below:

PBX Equipment Floor Loads for a 500,000-Square-Foot Tower

Functional Group	Typical Floor Load (pounds/square foot)
Frame room	100
PBX switch and associated electronics	100
Power supply, batteries, and air conditioning	300
Maintenance facilities, etc.	100

Electric Power

The 3,000-line PBX system will require approximately 40 kilowatts of AC power; 208 VAC, single-phase power is generally required by the equipment. The standby battery power supply, which typically operates as an uninterruptible power supply (UPS), converts the 208 VAC power to 48 VDC to operate the PBX electronics. UPS systems for PBX operation typically have a full-load duration of two hours when primary AC power fails.

Ventilation—HVAC

OSHA and most city codes require a complete change of atmosphere in the PBX power supply room at least four times each hour. The corrosive fumes emitted by storage batteries should be considered in the design of power supply room ducting, and all ducting should exhaust outside the building.

The 3,000-line PBX systems discussed above will dissipate approximately 125,000 BTU/hour above the ambient level, and this substantial load is concentrated in the room containing the PBX switch and associate electronics. The PBX system environment should be held to a temperature of approximately 70°F—plus or minus 10°—and 50 percent—plus or minus 20 percent—relative humidity (noncondensing). Installation of a Halon fire control system should be considered.

COMMUNICATIONS CLOSETS AND RISERS

Closets

Communications closets are usually provided as small rooms, aligned vertically throughout the building core to permit vertical riser cable installation and tapoff or distribution connections. The size of these closets is determined by the amount of equipment to be installed and the decision as to whether "building space" or "tenant space" is to provide for the equipment.

Figures 11–1 and 11–2 show two general types of closets that serve most ordinary telecommunication needs. Further details are shown in Figures 11–3 and 11–4, where the vertical risers are enclosed and/or exposed throughout the building.

It is strongly recommended that the communications closets be of sufficient size to allow parallel operations of a local telephone company cabling and a private communications concessionaire cabling. For a building of this size, assuming a 35-story building is involved, a 50-square-foot communications closet should be sufficient for the mounting of patch panels, cutdown blocks, and occasional mux/demux equipment, and power supplies. Communications closets should be vertically aligned throughout the building, as closely as feasible, and should be equipped with power and light.

Vertical Riser Cables

Assuming that the building PBX system will operate 1,500 telephones within the tower itself and that the local telephone company will furnish private-line services for 1,000 more, we can estimate the number of telephone risers likely to be installed in the tower. Taking the PBX system first, the

FIGURE 11–1
Walk-In Closet

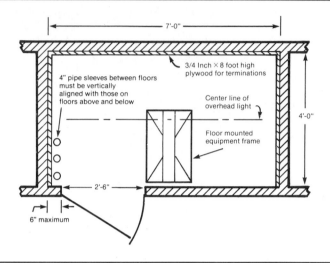

Note: Typical for 9,000–10,000 square feet of usable floor area.

FIGURE 11–2
Shallow Closet (wall-mounted equipment)

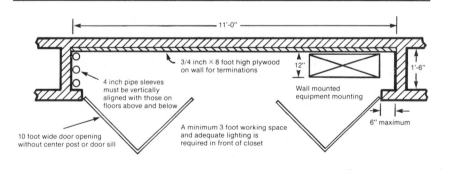

Note: Typical for 9,000–10,000 square feet of usable floor area.

FIGURE 11–3
Closed Riser System

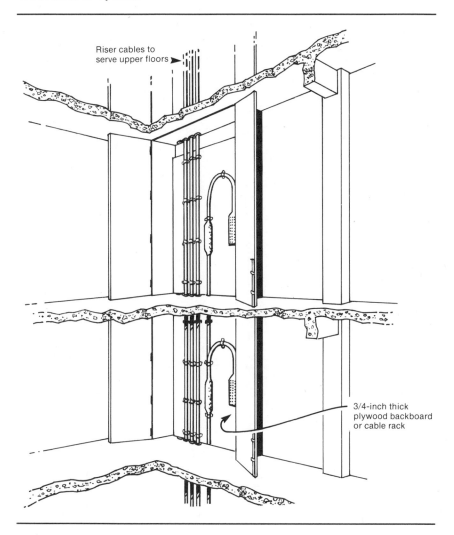

Riser cables to serve upper floors ▸

3/4-inch thick plywood backboard or cable rack

1,500 telephone lines can be assumed to be equally distributed through the 35 floors, so that approximately 45 PBX lines would serve each floor. Each PBX line requires three pairs of conductors, so that each floor would be wired for approximately 150 PBX line pairs.

The local telephone company private-line service is a bit more difficult

FIGURE 11–4
Open Riser System

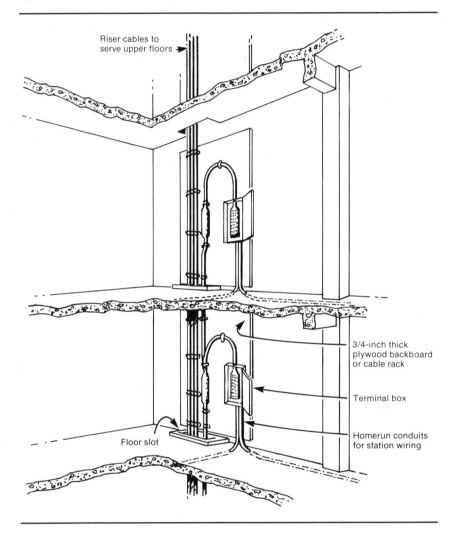

Riser cables to serve upper floors

3/4-inch thick plywood backboard or cable rack

Terminal box

Homerun conduits for station wiring

Floor slot

to predict because different types of local telephone company facilities may be provided, and these will affect the numbers of line pairs required. Assuming that the 1,000 local telephone company telephone lines are equally distributed through the 35 floors, approximately 30 local telephone company lines would serve each floor. Each local telephone company line may require as few as one or (as in four-wire services) two pairs of wires. Assuming the latter, each floor would be wired for approximately 75 pairs of local telephone company lines.

Based on the foregoing, we would expect to see a total of 5,250 **PBX** line pairs and 2,625 local telephone company line pairs entering the lowest communications closet from the basement to begin the vertical run. As noted above, a 4-inch ID conduit, or riser sleeve, can accommodate up to a 3,600 line pair cable, with a 2,700 line pair cable being the more common maximum size. The lowest floor communications closet would therefore require four 4-inch ID risers for PBX and local telephone company cables, plus others for future service expansion. It seems reasonable to assume that a total of eight 4-inch risers would be needed in each communications closet for the first 10 building floors, with the number declining as telephone riser line pairs decline in upper floors. A suggested arrangement is tabulated below:

Five Hundred Thousand-Square-Foot Tower Communications Closet Riser Distribution

Floors	PBX Pairs/Closet Capacity	Local Telephone Company Pairs/Closet Capacity	Four-Inch Risers/Closet
1– 7	5,300–4,000	3,500–2,600	8–6
7–15	4,000–2,700	2,600–1,900	6–4
16–24	2,700–1,400	1,900–1,200	4–2
24–35	1,400–0	1,200–0	2

Accessibility

Conversations with telephone company planners indicate that communications closet accessibility is at least as great a problem to them as riser capacity. Telephone company personnel frequently find that pneumatic sensor lines for environmental controls, dedicated security system cables, and similar facilities occupy riser space or closet wall space (or both) to the exclusion of telephone cables. In extreme cases two or three fragile copper pneumatic lines, only a fraction of an inch in diameter, occupy a 4-inch riser sleeve and exclude its use by other services. It is recommended that anticipated communications closet uses be reviewed to optimize the layout of specially dedicated facilities for security, environmental control, and other factors.

ROOFTOP FACILITIES

Among the many requirements of the interior cabling and wiring is the provision of access to the roof level to feed signals to and receive signals from satellite and microwave facilities. The communications closet riser line should extend to the roof level and/or to the level at which communications equipment will be mounted. The building should have provisions for private-leased rooftop facilities. Accordingly, a multiplicity of conduit entries to the equipment area should be provided. In general, a 2-inch conduit would suffice for two-way radio and/or a microwave facility, whereas a 4-inch conduit may be required for ease in running a waveguide connection from the rooftop level down to an equipment room that could be located immediately below. In all of the planning, sufficient sleeves or knockout areas should be provided to allow easy access to the equipment room and/or to the roof area.

Microwave and Two-Way Radio Facilities

A microwave transmit/receive antenna is generally a parabolic-shape "dish" varying in diameter from 2 to 10 feet. Most rooftops have sufficient space to mount these microwave relay antennas. The major criterion is clearance of the horizontal path, since the transmission is line of sight. Microwave frequencies available for business communications purposes range between 2 GHz and 23 GHz, and the higher frequencies use the smaller antennas.

Main considerations for installation of these antennas, once line of sight has been cleared, are frequency coordination, FCC licensing, and assurance that no non-ionizing hazard exists. Once the frequency coordination and licensing process has been cleared, hazard to human beings can be minimized by mounting the equipment so that the maximum power density level is at inaccessible locations. This could mean mounting the dish on a parapet facing away from the accessible areas or on a penthouse wall or tower so that access is impossible.

Two-way radio antennas, or "whip" antennas can be mounted at many locations on the rooftop. Once again mechanical clearance is the most important factor, since these antennas radiate fairly omnidirectional signals and do not require line of sight to a receiver. In all cases antennas mounted on the roof should be carefully grounded to protect them and the building from lightning damage. Weatherproof connections should be made to bring the antenna cable to the equipment located either in adjacent shelters or on the floor below.

If a tower is considered for the mounting of two-way radio antennas and/ or microwave antennas, consideration should also be given to the need for any aircraft obstruction lighting, a matter totally dependent on the building height and its proximity to airports.

SHARED TENANT SERVICES

There are many pros and cons to the "sharing" of services in one building or building complex among the many tenants. New telecommunications technologies make it possible to share telephone and data switching equipment and computer and processing equipment in a cost-effective manner. In many cases the variation in usage patterns can provide very meaningful cost benefits.

This sharing is possible with the maintenance of the appearance of an individual service, with full security and privacy afforded to each user. Unfortunately, in many cases it is very difficult to convince the user of the degree of security that has been provided.

New computer-controlled PBX systems are available with complete compartmentalization that provides the same privacy in shared switching that would be available in single-purpose, customer premise equipment. There is a physical and economic "pro" in using such a shared service, but there is also a psychological "con" present regarding privacy and security of information.

All of this must be carefully considered in analyzing the method of marketing a shared service. Careful consideration must be given to the fact that a large installation reduces unit cost as well as operation and maintenance costs. Larger installations of the more sophisticated computer-controlled switches also offer a wider range of features and greater flexibility than most smaller units.

Aggregation of communication costs through the sharing of voice and data trunks is another approach to sharing that minimizes trunking, minimizes turnkey installations, and can be a major cost-saving element.

Sharing of computers, data bases and word processing systems has been represented as another potential for major cost savings. This has proved to be of value in the case of multistation access to a mainframe and access to data base nodes, but not in the case of word processing.

Word processing systems proliferate at such a wide variety of prices and with so many competing features that centralization has not proved economical. Multistation facilities within an office, with automatic paper feed printers and high-speed draft printers located on the same floor have become the norm. A central facility serving local printers and local terminals does not provide sufficient economic advantage for serious consideration.

However, computers can be compartmentalized in the same fashion as computerized digital PBX equipment and provide the same degree of security. Having available the capability of large mass storage and high-speed processing is advantageous to many types of business. However, there is also the general feeling of insecurity and a reluctance to store sensitive data in devices not fully under the tenant's control.

Sharing of data bases, however, does appear to be advantageous. The multiple data bases serving U.S. industry generally have nodes in key cities, and access simply requires a local call. The aggregation of such data bases and nodes within the building complex can provide for cost saving and time saving and should be carefully explored.

MULTIPLEXING

A communications channel exists in at least two dimensions—it exists over a period of time, and it has breadth, or bandwidth—its extension in the frequency domain. Additionally, the channel may have spatial characteristics and exist only between distinct points, as in the case of a cable connecting one telephone with another, or a directional antenna projecting radio communications to a favored location. Channels can sometimes be characterized by propagating electromagnetic energy in particular planes or modes—vertical, horizontal, circular—and be distinguished from other channels sharing the same frequency, temporal, and spatial dimensions by virtue of their polarization characteristics.

The two basic dimensions, time and frequency, are often exploited to permit a channel to be shared among several data sources, although the other channel attributes can also be used in the same way. We are most familiar with frequency multiplexing, the division of a channel's bandwidth into subchannels, each having distinct frequency limits or boundaries. Frequency multiplexing is employed to permit the broadcast radio and TV bands to carry the programs of many stations simultaneously; one's radio or TV receiver is tuned to a particular subchannel or frequency to separate the desired signal from all others sharing the same frequency band. The entire electromagnetic spectrum from long-wave radio to optical radiation can be considered as a continuum, and the hundreds of thousands of signals existing therein coexist by virtue of the distinctness of their carrier frequencies or channel occupancy, otherwise known as frequency multiplexing. Telephone trunks often employ frequency multiplexing to combine individual voice signals. The trunk cable has much more bandwidth than is required for any one conversation. Each signal is modulated onto a distinct carrier frequency in a manner similar to that employed by radio stations to place program material on a carrier for radio transmission. Many voice signals are "stacked" in frequency by using closely spaced carriers. The ensemble of carriers and signals is called a group, and groups of groups (supergroups) are sent over the trunk to distant central offices, where the signals are separated by carrier frequency and demodulated before sending over local loops to individual telephones.

Time multiplexing is also a common human experience. We often do several things at once, or carry on two conversations at the same time. We time-share our attention, communicating now with one thing or person and then with another. This is time multiplexing. In simultaneous conversations one time shares aural communications channels—voice and hearing—among several participants. The meaning of each interchange is digested as though attention were individually focused, and there is the *illusion* of simultaneous communication because of dealing with more than one situation in the same span of time. Note that "the same span of time" is not "at the same time," or truly simultaneously. Similarly, one's telephone conversation is sometimes carried over the same trunk cable as those of others by use of time multiplexing. Each voice channel is sampled for an instant and sent, in sequence, with samples from other local voice channels. At the other end of the cable the time-sequential samples are separated, or demultiplexed and reassembled on individual local loops in the recipient's city. If the sampling is done often enough, the reassembled signal contains all of the information sent, and the listener perceives that the connection (or communication channel) with the sender is continuous and exclusive. The fact that rapid sampling of information can substitute for continuous connection is of enormous importance in communications; it was first quantified by Harry Nyquist of Bell Laboratories in the 1920s and is known as the Nyquist sampling theorem.

Spatial multiplexing is also evident in our everyday lives. Home telephones are connected to a central office by means of an individual wire pair. Our wires may cross or be bundled with those of others without interaction because the communications channels represented by each wire pair are spatially separated and distinct by virtue of the insulation covering each wire. Spatial multiplexing also occurs when two radio stations share the same frequency but are separated geographically or transmit from the same location in different directions.

Designers of high tech buildings employ various multiplexing methods, sometimes in combination. Multiplexing allows voice, video, and data to coexist on the same communications channel and permits exploitation of the available channels to the greatest economic advantage.

LOCAL AREA NETWORKS

Means to interconnect data processing equipment within the confines of an office, hospital, factory building, or building complex are called local area networks, or LANs. Such networks provide resource sharing and communications among these devices at high speeds and are typically owned by the user. The conjunction of high speed and low cost is not contradictory because the transmission paths are short. The paths vary from several meters to several kilometers in length, short enough so that even ordinary twisted pair telephone wiring can support reasonably wide bandwidths, such as 56 kbps. LANs have become important because of the proliferation of the personal computer and its ancillary devices. The mainframe computer of the 1950s became the several minicomputers of the 1960s and 1970s; these, in turn, have been replaced by the microprocessor-based personal computers that inhabit many desks, from executive to clerk. The nature of office transactions has not changed, however. Most of the information generated by an individual is intended for someone within the same office or complex; estimates range up to 80 percent internal distribution. However, now that there are many information-generating devices, computers, and related equipment, the distribution means must accommodate this proliferation.

Direct lines suitable for interconnection of a large computer and its terminals and printers are impractical as a way to interconnect many PCs, since interconnections grow geometrically with the number of PCs. The intraoffice private-branch exchange (PBX) that serves all desks with telephone service is not a good alternative. The data speeds supported by the PBX and its lines are too low in many instances, and several lines in parallel would be required. In addition, the nature of the traffic is different; data communications are conducted in bursts, with relatively long pauses in between. To efficiently use the interconnection system requires that the communications path be relinquished after each burst so that the facilities may be used by others; but the PBX requires several seconds to establish a connection (the dialup time). In contrast, voice conversations are protracted and relatively infrequent; the dialup time is inconsequential and tolerable. Since dialup delay is often longer than the data burst, the use of a current generation PBX and its lines is not practical and would not work with standard data transmission protocols, however tempting, because they are in place, well accepted, and often paid for. A further drawback concerns the network configuration that use of the PBX imposes; all lines lead to a central location—the PBX equipment. If this malfunctions, the entire interconnection system is inoperative.

Local area networks avoid the disadvantages of interconnection through the intraoffice telephone system by offering a communications channel tailored to the special needs of efficient data communications. LANs provide high-speed data transfer, there is low overhead time for establishing interconnections, and they can be configured to avoid centralization and catastrophic failures. Like telephones, most LAN wiring is inexpensive. Careful planning to accommodate future changes and expansion of the system is not needed, and the wiring can be installed everywhere at little additional cost when the office or building is being outfitted.

There are three ways to characterize LANs: by the network layout (topology), by the means by which a device gains access to the network, and by the interconnection medium (wire or cable) and its data capacity. Confusion can result because one characterization does not imply or preclude another, and all three are needed to fully define a LAN. To add to the confusion, standards are slow to emerge, and LANs are further distinguished as general purpose, special purpose, or proprietary—depending on their ability to interconnect any device, a class of devices, or only devices of one manufacturer.

Network Topology

The network topology is partly determined by the type of control intended for the LAN. If the function of one device to be netted is dominant, then control should be centralized at that device, and it becomes the master node of the LAN; the master sets the order of "talking" on the LAN and polls the other nodes to find their need to send information. If many devices have equal importance, control should be distributed, and there is no master node; becoming the talker is set by more democratic means, as will be described below. Networks having a master node can effectively use star or tree network configurations of wiring. The master is located at the center of the star or base of the tree. A choice between these two might be dictated by the structure of the building and the ease with which wiring is installed. Ring and bus topologies are suited for distributed-control LANs. There is no dominant node for these configurations. The bus configuration is easily expanded and is tolerant of failures in devices attached to the bus. If distances between devices on the LAN are large, the ring configuration is favored. Each device receives data from one side and reconstitutes it before sending it onto the other side; in this manner, weak signals are constantly being restored. The ring has several disadvantages; failure at any node brings the entire network down; there are delays associated with regenerating the signals; and the devices are more complex and costly because of their active role in retransmission on the network.

Access Methods

There are various methods of network access by which the device at a LAN node takes its turn to transmit on the network and send its data to any or all other nodes. For LANs having master stations, access is usually granted by the master. Each node is polled; if the device at a polled node has data to send, it so indicates, and the master grants access and allows the device to talk. Access is typically granted for a fixed, short time. The device might have much data to send, but it is not permitted to tie up the network until its data are exhausted; rather, the device must wait its turn and send data in many short segments. Distributed-control LANs can grant access by fixed schemes or by random techniques. Among the fixed methods is token passing. A unique data word, the token, is passed from node to node. If the device at a node has data to send, it does so before passing the token onto the next node. An alternative method is the slotted ring; a period of time is divided into segments called slots, and any device having data to send can do so in an empty slot. The device to which the data are addressed "empties" the slot and indicates that it is available by inserting a data word in the slot to indicate that the contents are no longer needed.

If the time required to poll, pass tokens, or wait for empty slots cannot be tolerated, an access scheme called carrier sense multiple access with collision detect (CSMA/CD) is often used. There is no order of access, and any device having data to send can do so if the network is free at the moment. Devices sense carrier energy on the line and refrain from sending data if carrier is present; this is called listen-before-talking discipline and is how most of us conduct ourselves in a group. If two or more devices send data simultaneously because both sense a free line at the same time, data collisions occur, and messages are garbled. Each device's carrier sense circuit now notes excess carrier energy on the line and stops sending its data promptly. After a delay, made different for each device on the LAN, the device again determines if the line is free and then resends its data. With CSMA/CD there is no determined order of access and no minimum waiting time. Devices with more data to send obtain line access more often, and the distribution of access is fairly weighted in accordance with the transmission needs of each node.

Interconnection Media

LANs can be characterized by the amount of data they can carry and the speed of doing so. There are two types of networks under this classification: broadband and baseband. Broadband LANs employ frequency-division multiplexing to partition the spectrum afforded by coaxial or optical fibre cables into a number of channels. Cables are suitable for broadband networks because they can support a wide range of frequencies. Used in the broadband mode, coaxial cable usually has more capacity than is required for the LAN. Some channels can be used for the LAN function, and others can be used for distribution of voice and video or for energy and security monitoring. The broadband network is similar to that used in cable TV systems and shares some of the low-cost hardware used in CATV practice. Optical cables have very large bandwidths and are suited for multifunction networks and where size and weight of the line are at a premium. Baseband LANs do not divide the spectrum into channels and use the "base" of the frequency band in its entirety for all the data traffic, usually over twisted-pair wiring. Time-division multiplexing must be used so that all nodes can share this single channel; other services cannot share the baseband network. Twisted pair wiring is very low-cost but has low bandwidth and high loss. It is suited for baseband networks that use the ring configuration so that regeneration at each node overcomes the wire's losses. Because of its low-loss properties, coaxial cable is also suited for baseband networks in which long runs between nodes are encountered and in which signal regeneration is not used.

BYPASS

Means of avoiding the use of the conventional telephone network are called bypass technologies. Bypass communications may be an alternative to the local loop between the customer and the telephone exchange, or it may avoid the use of the long-distance telephone network; in some cases, avoiding both is effected. Telephone companies (AT&T, the BOCs, and independents) have responded by offering bypass services themselves and have marketed high-capacity communications alternatives to their own traditional wire-line facilities that have been in place for decades. Bypass facilities can be owned, therefore, by any interested party: the user or user's designate (private systems), groups of users (shared systems and multi-tenant systems), radio common carriers, and wireline common carriers (the telcos that are otherwise bypassed).

Bypass technologies include microwave radio, satellite radio, coaxial cable, and optical fiber cable. Bypass is employed to reduce costs, to provide services difficult to acquire from conventional telco offerings, or most often, a combination of these two reasons. The most common example of the first reason is the use of another long-distance service for home or business, such as MCI or Sprint. Examples of the second reason include video teleconferencing and high-speed data communications. Several types of bypass services are discussed in the paragraphs that follow; both local-loop and long-lines bypass techniques are presented.

DIGITAL TERMINATION AND ELECTRONIC MESSAGE SYSTEMS (DTS/DEMS)

In 1978 the Xerox Corporation petitioned the FCC to allocate a portion of the radio spectrum at 10 GHz (commonly called X band) for a common-carrier broadband network service; Xerox had named its new proposed service XTEN. The FCC responded positively in 1981 with a Report & Order that reserved 130 MHz of spectrum at 10.55 to 10.68 GHz for digital termination systems (DTS) and the digital electronic message service (DEMS), both of which were modeled on the XTEN concept.

Why DTS?

The new service is designed to provide business users with data transmission capacity for intracity communications that are not easily obtained from the local telephone company. In a sense, DTS/DEMS is not a bypass service, since it is not an alternative to an existing, readily available telco offering. Traditional local loops (the telephone line from the customer's premises to the telephone central office) and central office switching equipment are not geared to handle high-speed, high-quality digital data traffic. The service does present a bypass alternative to wideband private lines that can be obtained from the telephone company; these lines are not part of a switched service, however, and the user must have fixed destinations in mind when contemplating their use. DTS is a local distribution service that carries data and other wideband traffic, such as high-speed facsimile and video teleconferencing, between users and central stations called nodes by means of point-to-point or point-to-multipoint full-duplex radio. Traffic can be switched at the nodes, allowing users to treat the system much as the telephone or postal system. Intensive reuse of each operator's radio channel is encouraged by the FCC, and extensive intracity networks can be achieved as in the case of cellular radio for vehicular telephones. Data rates from 75 bps to 1.544 Mbps (T–1 service) can be carried by using equipment now being offered by several manufacturers. Interconnection of DTS nodes is provided by radio channels in the DTS radio band dedicated to intranode traffic. Connections among cities are effected by use of other services—such as long-haul microwave, satellites, and coaxial or fiber optic cables. A DEMS network consists of interconnected DTS nodes, and the DTS links between users and their local node form the local-loop portion of the service. The nodal stations are analogous to the telephone central office, or switching exchange, and the interconnections of the node correspond to AT&T long-lines circuits. DEMS license applicants proposing to build in 30 or more cities can use 5-MHz channels between users and nodes and are classified by the FCC as extended carriers. Others, applying for networks involving 1 to 29 cities, are assigned 2.5-MHz channels and are considered local carriers.

In its first Report & Order, the FCC recognized that 130 MHz of spectrum might be inadequate and that the 18-GHz region would be examined for additional DTS/DEMS allocations. It was also noted that private carriers might be permitted to use the service. A Notice of Inquiry by the commission resulted in the second DEMS Report & Order in 1983 that authorized the service at 18 GHz and modified the rules at 10 GHz. The commission did indeed permit private carriers to operate DTS systems and DEMS networks; further, the distinction between local and extended carriers was removed by allowing all to use 5-MHz channels for user-to-node and node-to-user communications.

The DTS "Local Loop"

Currently, there are two methods of configuring DTS "local loops" between the node and the user; each has hardware available from one or more vendors. The first method is point-to-point radio, for which a separate communications path to each user is established by means of highly directional antennas at both the user and node ends of the path. This method is attractive to private users that have known, specific locations that are to be included in a digital communications network; the cost of such a configuration is relatively low because users have common goals and communications requirements. An implication of this method is that the network protocols can be restrictive; node hardware and software do not need to accommodate service on demand as is necessary with common-carrier systems that cater to many independent user-customers who cannot be expected to coordinate with one another in gaining access to the network. The switching tasks required of the node can be greatly reduced and, in some cases, users can be serviced with private, dedicated lines, and no switching is needed; the node acts solely as a radio station and no value-added services are performed.

The second DTS local loop configuration method involves point-to-multipoint radio, for which the node operates in a manner similar to that of a broadcast station. This approach is suited for common carriers, since they have to serve system customers on a first-come, first-served basis and often do not know where user stations will be located at the time the system's node station is sited and constructed. The node can transmit data to users omnidirectionally, and all user sites within the reliable transmission radius of the node can be serviced. Each user-receiver "listens" to the transmission and recognizes messages intended for it by means of a unique address sent prior to the body of the message. More typically, the node transmits simultaneously in three or four sectors, each covering 120 or 90 degrees of azimuth about the node's geographic location. In this manner the system operator can "reuse" the channel assigned to him/her by properly assigning subchannel frequencies and polarizations within each node and creating a network of nodes, all utilizing the same channel. Such systems combine time-division and frequency-division multiplexing (TDM and FDM) that were explained earlier in this chapter and in Chapter 2 as approaches to sending multiple signals over a channel. The operator divides the channel of 5-MHz bandwidth into three or four subchannels to afford sector coverage about each node (FDM) and then sends different message batches simultaneously in the sectors to reach users known to be located in each of the various sectors. Users wait until the message stream they are capable of receiving contains data addressed to them, at which time the message is extracted from the stream (TDM).

User communications are arranged so that each of many users appears to be continuously on-line, or served by the node. This is possible because the typical rate at which a user can send or accept data is far less than the communications rate over the DTS channel. A typical high-speed user terminal (digitized voice, fast fax, or computer port) will send or receive at 64 kbps; a typical DTS subchannel can support 24 such users simultaneously. Each user's data stream is broken into packets containing several hundred information characters (text or numerical, usually). A packet might contain destination and source addresses and error control data. The aggregate of all noninformation data associated with each packet is called overhead; overhead burdens range from 1 to 10 percent of the total packet. There are several standards that prescribe packet format, and system operators can usually handle one or more of these to meet user requirements. A packet containing 250 characters takes only 1.3 milliseconds to send or receive, and each of 24 users is served with a packet about 32 times per second. A data buffer at each user terminal accepts or prepares the packet for communication at the DTS channel rate, about 1,600 kbps, and presents the data to the user at a 64 kbps rate. From the system operator's viewpoint, the packets are small chunks of disconnected data, each user getting only 1 out of each 24. From the user's vantage, the buffers give the appearance of undivided attention being paid to each user, whereas each user is really waiting about 96 percent of the time.

Packets for a particular user can be assigned a fixed time slot in the frame of 24 allotted for the users. Information is sent to identify the beginning of each frame, and users need not look for their addresses. They always know that the contents of the slot are theirs; this is called fixed-assignment TDM and is a private-line local loop. The slot is always there and the user pays accordingly. Alternatively, assigning slots as the user requires is called demand assignment multiple access (DAMA). The DAMA user is not always occupying a slot and therefore pays less; the operator compensates by making the same number of slots available to more customer/users on an as-required basis.

Either of these time-division multiplexing techniques can be extended in both directions, serving more users at lower rates, and even intermixing data rates. One DTS channel can be used to serve a variety of users, each with widely differing terminals and data rates. The buffer at each terminal is tailored to standardize the data as seen from either the DTS provider's or the user's viewpoint, resulting in great system flexibility.

User-Node Communication

So far we have shown how users receive messages. In order to originate traffic over a DTS system operated by a common carrier, the user employs techniques similar to those used to send a message on a LAN system. If a user has a fixed assignment, there is an allotted time slot in the user-to-node time frame, and permission does not have to be requested or granted to send data to the node. In the more flexible cases, such as DAMA, a portion of the user-to-node time frame is reserved for requests for service. Either users contend for attention during this time, or they are assigned short request slots and make their need for origination service known to the node. The node, listening for collision-free requests, sends grants to users that inform them of the slots reserved for them in the next or subsequent frames. If request collisions occur among users, grants will not be sent by the node, and the users will try again after several frames have been transacted without acknowledgment of their service requests. Sophisticated system protocols will accept and grant requests for service that are rate dependent. A user who has an immediate demand for a high volume of traffic might include in a request, along with identification, an indication of the amount of data to be sent to the node. The node can then grant several slots in the frame or extend slot time and delete others in the frame if other demands for service are low at the moment. The same flexibility can be employed in the node-to-user direction, but is more straightforward than user-to-node to implement, since the request/acknowledge cycle does not have to be transacted. The node can simply pass the incoming volume of data on to the user without further ado, assuming that the user terminal buffer has been sized to accept the largest possible expected burst of data traffic. If the terminal buffer is too small, the node will have to buffer the traffic and send at a lower rate or adjust the user rate in accordance with the amount of undigested traffic in the user's buffer by means of a reverse request/acknowledge transaction.

The Node Station Rooftop Antenna

The radio communications path between each user and the node station must be line of sight; that is, the user and node antennas must be visible, one from the other. This is because, at the high radio frequencies involved, radio propagation is similar to that of visible light, and terrain or building blockage of the path defeats communications. The node station antenna is situated on the roof of one of the highest buildings in the city, or on a high promontory affording a good view of the business district. The node station location is selected with this ability to "see" as many user sites as possible. Of course, node stations for private DTS/DEMS networks may be sited to communicate with known user stations and might not be optimum from the standpoint of a common carrier who should be "visible" to as many potential user/customers as possible. The user antenna is typically located on the roof of the user's building and is usually a 24-inch diameter parabolic dish antenna that is pointed at the node station. The user antenna is used for both incoming and outgoing traffic and can handle any and all data rates or system protocols. The antenna is associated with a radio transmitter/receiver unit, usually located on the roof with the antenna. This unit converts the radio signal to a much lower frequency that is capable of being conducted to the user office by means of low-cost coaxial cable, such as used in cable-TV installations. This cable is only one quarter of an inch in diameter and is easily routed through the building's cable or piping ducts. One antenna, radio unit, and coaxial cable can serve more than one user; each attaches to the cable in a manner similar to that used in LAN networks. Users need not have any communications goals or requirements in common other than that they reside in the same building and wish to reach the same node station. Data rates, usage patterns, types of traffic—all may differ and yet share the same inexpensive cable routed from the roof equipment throughout the building. Users have their own separate terminal equipment that sends, accepts, and buffers their traffic. Access to the node is requested and acknowledged independently of other users, as though each were located in separate buildings and did not share a common antenna, radio, and cable.

MULTIPLE-ADDRESS RADIO

In contrast to the high data capacity of the DTS/DEMS service, the multiple-address radio service (MAS) is a narrow-band alternative available at much lower cost. The FCC was petitioned by the utilities industry for a spectrum allocation to permit remote meter reading and energy load management; the petition was granted in 1981. In early 1982 the FCC released a further Report & Order that provided 14 paired channels and 8 unpaired channels at 900 MHz for other private users (noncommon carriers) to conduct two-way and one-way general business and government-related traffic in addition to the channels set aside for the power industry. This service is similar to DTS in that operations are permitted in the point-to-multipoint node. In MAS radio the concept of the DTS node station becomes that of the master station. In both cases, omnidirectional transmissions emanate from the central station, but the master station is most often at the place of business from which the communications originates and at which responding traffic is to be used. In contrast, the DTS node is only a central gathering point for traffic that is then passed on to other nodes for ultimate distribution to other users, and no traffic originates at the node.

Propagation of radio signals at 900 MHz is quite different from that at 10,500 MHz, at which DTS signals are allocated. The MAS channels are slightly above the UHF TV band and share TV's ability to provide nonoptical (not line-of-sight) communications paths that penetrate buildings to a limited extent. User stations' antennas can be small (typically 4 by 12 inches) and inexpensive ($100 or less); often they can be mounted near a window in the office to be served, even if located on a low floor of a high-rise building among other similar structures. User radios are also small (less than 1 cubic foot) and can be located in the user's office; some have built-in modems that directly interface with telecopiers, computers, or other equipment that has an RS–232 dataport. The master station antenna is best located on the roof of a high building, preferably near the center of the area to be served.

Communications at rates up to 9600 baud can be supported over the 25-kHz MAS channels. Two types of services seem to predominate among MAS license applicants. In one case a central office has data to disseminate that are of interest to its outlying offices; if the data are stock quotations or other general information requiring no interaction with the originator, a one-way channel can be used. In the other case, a central office may wish to monitor conditions at a number of locations or may wish to collect data from them. The master station polls each of the remotes and each, in turn, responds over the remote-to-master paired channel. This is the operating mode for meter reading and load monitoring and is popular for remote security alarm monitoring by such firms as ADT.

■   CONCLUSION   ■

With many different transmission facilities available and with the continuing rapid pace of change, no one technology is the panacea in providing high tech telecommunications for a building. However, a combination of some or all of the transmission facilities discussed can both protect the developer from obsolescence and provide a cost-effective and profitable approach for services to the building and its tenants.

Authors

Joseph L. Stern
President
Stern Telecommunications
New York, New York

With more than 40 years of experience in electronics and communications, Joseph Stern is a nationally recognized authority on broadband communications and emerging technology. He and his firm are actively engaged in consulting for equipment manufacturers, government agencies, cable television operators, investment houses, financial institutions, architects, real estate developers, and specialized common carriers. His experience ranges from the basic design of complex business telephone systems, the design and construction of cross-country microwave systems, 15 television stations, the development of a stock exchange communication system, the development of fiber optics local area networks, to the development of the unique Metropolitan Transmission Center for interconnecting broadband communication systems. He has also been in charge of the engineering efforts undertaken in developing the United Satellite Communications, Inc. DBS system. Stern is a regular contributor to engineering handbooks and is a frequent lecturer on the future of communications before engineering societies, industry conventions, and special business communications seminars.

Prior to founding Stern Telecommunications Corporation, Stern was vice president and director of engineering at Goldmark Communications Corporation, a company he helped found in 1972. Before that, he held the post of vice president of engineering for CBS Television Services Division; and for 26 years he held various executive engineering positions at CBS. While there, he had primary responsibilities in broadcasting, cable TV, and facsimile and advanced communications technologies and applications. Prior to joining CBS he held engineering posts at RCA, Melpar, and in the Army Signal Corps, where his primary activity was the design of radio frequency transmission equipment and antenna systems.

Stern received his B.S. in electrical engineering from the University of Connecticut and did graduate work at several other schools. He was the recipient of the first University of Connecticut Annual Engineering Achievement Award.

Robert E. Weiblen
Integrated Systems Planning
Baltimore, Maryland

Robert E. Weiblen has been a practicing communications engineer for 29 years. Recently, he has been concerned with applications of DEMS and UHF radio to business data communications and wireline bypass. He has filed FCC license applications for pay-TV stations, cellular radio, direct broadcast satellites, and business communications services via domestic and international satellite facilities. Weiblen has developed computer-based microwave data links and has devised programs and algorithms for communications systems analysis. He is a cofounder of Household Data Services, Inc., in Reston, Virginia, and currently heads his own consulting firm in Baltimore. Weiblen holds degrees from Stevens Institute of Technology and Columbia University; he is a licensed professional engineer and is active in several FCC standards committees.

T. J. Theodosios
Integrated Systems Planning
Baltimore, Maryland

T. J. Theodosios has more than 22 years of professional management and technical experience in telecommunications and data processing. He started his career in the Air Force at Cape Canaveral, after which he joined NASA, supporting the Gemini and Apollo programs, designing and implementing audio, video, and R.F. communications systems. In 1969 he was project manager for all ground support communications at the Kennedy Space Center for our first moon landing launch. Later, Theodosios spent five years developing systems related to data processing for a major regional hospital in Baltimore. He was also on the faculties of both the University of Baltimore and Loyola College as an instructor of computer science and business administration. Theodosios then joined Commercial Credit Corporation, where he held increased levels of responsibility in telecommunications until his departure, at which time he was manager of telecommunications. In 1974 Theodosios implemented a 1,500-line ESS Centrex, the first in the state of Maryland. He managed the design and implementation of Commercial Credit's Plato system, an on-line educational system designed to be user-friendly.

His last position prior to forming ISP, Inc. was vice president of corporate telecommunications for Alexander and Alexander, Inc., where he created the telecommunications function and organization. During his tenure he also established the administrative services function, which included the purchasing, fleet management, and real estate departments. He was instrumental in implementing a national voice network at Alexander and Alexander, the first in the insurance brokerage industry. He developed a unique teiecommunications organization at A&A, which included multidiscipline experts in data and word processing, telecommunications, and strategic planning.

Theodosios received a B.S. in computer science from the University of Baltimore in 1972 and did postgraduate work at Morgan State University.

Chapter 12

Preparing Teleport Applications

Andrew D. Lipman
Pepper, Hamilton & Scheetz

Emilia L. Govan
Pepper, Hamilton & Scheetz

Outline

INTRODUCTION

Teleports are real estate developments in which the property owner or manager incorporates telecommunications and computer services and facilities (e.g., satellite dish antennas) primarily for long-haul voice and data transmission. The cost of these services and facilities can more efficiently be shared by a large set of users. By packaging access to these technologies, developers afford both their tenants and off-site users within a particular region flexible access to services that otherwise would be too costly or difficult to secure on an individual basis. In addition to on-site users, off-site users in numerous adjacent business districts can tap into the teleport by means of microwave and fiber optic cable links.

The teleport concept promises to provide cost efficient access for small- and medium-volume users who might not generate enough demand for satellite and other efficient transmission facilities, and for large-volume users who prefer having a turnkey operation rather than investing the time and effort necessary to develop a private telecommunications system. Teleport facilities equipped for multiple remote access also promise to solve problems for users in cities where existing telecommunications services prevent downtown construction and operation of additional microwave and satellite services or where available services do not satisfy particular needs.

The teleport facility will normally consist of several satellite transmit and receive stations providing cost-effective audio, video, and data communication services to and from subscribers. In addition to transmit-receive earth stations, a teleport may utilize terrestrial communications transmitting and receiving equipment, together with various support facilities, computers, and switching and connecting equipment.

A teleport site is normally located within, or close to, a downtown business center and is usually intended to serve the immediate metropolitan area as well as other portions of the surrounding region. The areas most suited to teleport development are generally those characterized by a rapidly expanding economy and those containing centers for energy, banking, finance, and related service industries. Such areas are likely to experience severe radio congestion and frequency coordination difficulties that frustrate the ability of businesses to utilize satellite and other communications services in a cost-effective manner. In these areas, teleports can meet the rapidly increasing need for efficient and cost-effective telecommunications to keep pace with industrial and commercial growth.

In addition to traditional data and voice offerings, a teleport can offer its users state-of-the-art communications services—such as videoconferencing and facsimile transmission capabilities—and make available high-speed, high-capacity business services in a digital format. Teleport facilities can be interconnected with various public switched and private networks and may be designed to access all domestic communications satellites, both existing and proposed, thus enabling users to communicate instantaneously with similar systems in other parts of the country. Teleport operators may desire to construct a terrestrial microwave network that will serve as the gathering system for communications accessing the teleport's satellite facilities. Microwave links can connect major user sites in the area directly to the teleport. A teleport may also use fiber optics transmission facilities to connect the teleport to various sites in the region.

The teleport will also likely offer access to international points. International earth stations that provide International Business Services (IBS) and international television services to area users can be located at the teleport. As a result of these and other arrangements, the teleport can serve as both a domestic and international communications gateway for the region in which it is located.

REGULATORY CONSIDERATIONS

Development of a real estate parcel for teleport services involves a number of regulatory considerations at the federal, state, and local level. While the teleport as a whole does not need to be licensed by the Federal Communications Commission (FCC), most of the individual communications facilities located on the teleport site, or otherwise associated with the teleport, do require various types of federal regulatory approvals. For example, earth station facilities and microwave facilities must be licensed by the FCC. Special requirements and selection procedures apply if the links are to be Digital Termination Systems, which are part of the Digital Electronic Message Service.

In addition to federal licensing requirements, other types of regulations may apply, including federal tariff regulation (e.g., for IBS earth stations). State and local approvals may also be required for certain types of facilities, for example, state certification and tariff and related regulatory requirements.

Teleport developers should therefore integrate legal and regulatory considerations as they conduct preliminary economic, marketing, and engineering analysis. The teleport operator will usually retain legal counsel as well as technical experts and other advisers in connection with the regulatory approval process.

The following discussion briefly describes some of the elements of FCC licensing applications for a few typical teleport facilities. However, this discussion is *not* intended to serve as a comprehensive listing of applicable requirements or as a substitute for a careful firsthand examination of the requirements and technical standards specified in the FCC rules. The main purpose of this chapter is to alert teleport operators to the type of information that must be developed or gathered, the type of resources that must be brought to bear on this effort, and the timing factors associated with the FCC licensing process. Failure to consider these and other regulatory issues could easily lead to unanticipated delays and other problems in implementing a teleport complex.

SATELLITE EARTH STATION APPLICATIONS

General Requirements

Each transmitting satellite earth station facility located at the teleport must be licensed by the FCC under Title III of the Communications Act of 1934, as amended, which governs the use of the radio frequency spectrum. Part 25 of the FCC rules applies to satellite communications, including earth stations. The application requirements will vary depending on whether the particular earth station is to be used for domestic or international communications. An example of a typical teleport application is attached hereto as an Appendix (Application of Gulf Teleport).

The application must be submitted by the party that will operate the station, and the "real party in interest" must be disclosed. That party will generally be either the teleport owner or operator or some other individual or entity leasing space at the teleport. To obtain a license for a domestic or international satellite earth station, whether C-band or Ku-band, an applicant must demonstrate legal, financial, and technical qualifications to construct and operate the facility.

The applicant must apply for a construction permit to construct or install the earth station and for a license authorizing it to operate the station. A single application may be filed requesting simultaneous authority to construct and operate the facility. Applications for simultaneous authority to construct and operate a domestic earth station must include an FCC Form 430 (dealing with the applicant's legal qualifications) and an FCC Form 403 (requesting authority to operate the station). The applicant may also request special temporary authorization (STA) to operate the station pending the processing and grant of the regular application.

To obtain authorization to apply for IBS and international television services, the applicant must submit FCC Forms 430 and 403 as well as an FCC Form 401 (for authority to construct the station). In addition, a separate application (commonly called a section 214 application) must be filed in order to obtain authority to operate as an international common carrier. The latter application can be filed simultaneously with the applicant's first earth station application. Alternatively, the section 214 application can be filed separately (prior to filing for any IBS earth station), as a request for blanket authority to serve as an overseas common carrier.

FCC staff will usually conduct a preliminary review of an earth station application shortly after it is filed. If it is found to be complete, the application will be placed on public notice for a 30-day period, during which time interested parties may comment on or protest the application. After the public comment period has expired, FCC staff will begin processing the application. In the absence of challenges by other parties, or of problems with the information submitted, the application should be granted within a few months of filing. Essentially the same application review process is followed for domestic and IBS earth stations.

It is also possible, in certain circumstances, to request an STA to operate the transmit earth station or stations. An STA request may be submitted with the initial application, or pending the filing of that application, when a regular application is filed or contemplated. An STA for a period of 60 days or less may be granted without the issuance of a public notice and the 30-day comment period. Applications for an STA are generally made by letter and should include a full description of the particulars of the proposed temporary operations, including frequency coordination data. STAs for domestic earth stations, which may normally be obtained in a matter of weeks, are normally not available for international earth stations. On the other hand, the processing time for IBS and international video earth stations has been considerably shorter than that for domestic stations.

Generally, earth station applications must contain a description of the proposed site where the station is to be installed, the exact geographic coordinates of that location, and a showing that the site is or will be available to the applicant. The technical characteristics and particulars of operation of the proposed station must also be described. If the antenna is larger than 30 feet in diameter, an environmental impact statement must be included. A frequency coordination and interference analysis report must be submitted, as well as a radiation hazard analysis.

Among the data required in connection with the legal showing are the names and addresses of the applicant's board of directors and of the shareholders holding 10 percent or more of the applicant's stock and a listing of other radio station licenses held. This portion of the application will also inquire whether the applicant is owned or controlled by a foreign company or government or whether any officer or director is an alien. The financial showing normally includes the applicant's recent balance sheet and may include any additional demonstration of financial capability (such as a bank letter of credit).

The application should also contain a statement demonstrating why the proposed facility would serve the public interest. Normally included in a statement are such items as the purposes of the earth station, the services it will provide, and the market it will serve. In this regard, it may be desirable for the applicant to provide a brief description of the planned teleport and a discussion of the ways in which the teleport will serve the public interest. If such a discussion is included in the first earth station application for the teleport site, it may simply be referenced in subsequent applications.

In preparing the application, an officer of the applicant must generally sign the various forms and provide the waiver required by section 304 of the Communications Act of 1934, as amended (waiving "any claim to use of any particular frequency or of the ether as against the regulatory power of the United States. . ."). Include the name and address of the person to whom correspondence relating to the application should be addressed. A technical certification must also be provided and should be signed by the person responsible for preparing the engineering data. The application's caption should indicate the city and state in which the proposed station is to be constructed.

Licenses for domestic earth stations are generally granted for a 10-year period. Construction of the earth station, however, must be completed within six months of the effective date of the authorization. If a construction permit alone is applied for, the applicant may specify the number of months required for construction, up to 18 months. In such cases, operating authority may be sought prior to completion of construction.

The authorization is automatically forfeited if the station is not ready for operation by the specified date. The station will be deemed ready for operation only after the applicant certifies in writing to the FCC that the station has been constructed exactly in accordance with the technical parameters and terms and conditions specified in the authorization. It is possible to apply for modification of the construction permit in order to obtain additional time to complete construction (on an FCC Form 701), but the application must be filed prior to the expiration date of the permit and must include a showing that failure to complete construction was due to factors not under the applicant's control.

Guidelines for Preparing Earth Station Applications

In conjunction with, or in addition to, the information called for in the required forms, the applicant for a domestic or international earth station should provide the following information:

Overall description. The description section should include a brief narrative of the earth station facilities and their proposed use, as well as the anticipated hours of use. If temporary authorization has already been obtained, that fact should be mentioned.

Specifically, this section should include the following:

Site availability.

Geographic coordinates (longitude/latitude to the seconds) and city, county, and state in which earth station is to be located.

Applicant's right to use site (by ownership, lease, or option to buy or lease).

Overall technical description.

> General description of facilities, operations, and communications services to be provided.
>
> Major structure and equipment, a functional block diagram of station equipment and scale sketch of site layout.
>
> Initial and potential communications capacity in terms of numbers and types of channels provided.
>
> Time required for construction.
>
> Estimated costs by major facilities component, such as land, building, transmitting and receiving equipment, antennas, channel equipment, engineering, and installation.

Points of communication.

> Space segment to be accessed by proposed station and other station with which it is to communicate.

Technical information. This portion of the application should include the following information:

Particulars of operation.

> Assigned frequencies (or transponder frequency plans where multiple r.f. carriers access the transponder).
>
> Polarization.
>
> Emission designations.
>
> Maximum main beam EIRP per carrier.
>
> EIRP densities per 4 kHz.
>
> Modulation characteristics and baseload configuration for each type of carrier.

Transmitting equipment.

> Number and type of transmitters.
>
> Output power and frequency tolerance of each type of transmitter.

Antenna facilities. For each antenna to be installed at the earth station, the following information should be provided:

> Brief description of antenna, including manufacturer, model, diameter, and type of feed.

Description of the antenna mounting, including range of the azimuth and elevation over which the antenna can be steered.

Antenna gain patterns.

A statement as to whether the antenna performance conforms to the standards specified in section 25.209 of the FCC's rules, dealing with two-degree orbital spacing of satellites.

Frequency range over which the station is capable of transmitting and receiving.

Transmit and/or receive main beam antenna gain and a 3 dB full beamwidth, specifying frequency at which each parameter is measured.

Polarization capabilities of the antenna.

Receiving system noise temperature, specifying the frequency at which each parameter is measured.

Elevation of antenna base above mean sea level; height of antenna center line above antenna base; and maximum antenna height above the antenna base.

There are also special technical showings that may apply to a particular facility. For example, an adjacent satellite interference analysis should be submitted with applications for use of small earth stations. Further, if an antenna's performance standards do not meet the specifications set forth in section 25.209(a) of the commission's rules, the applicant must demonstrate that the antenna may be operated in a manner that is consistent with 2° orbital spacing.

Frequency coordination. The application must also contain a frequency coordination study and interference analysis report demonstrating that the proposed operations will not create interference to other area carriers. In particular, the report must show that the frequency separation and local and natural shielding of the applicant's earth station transmitter will provide interference protection to nearby carriers (particularly terrestrial microwave carriers). Most applicants retain a telecommunications consulting firm to conduct the study, and the authors would generally recommend such a procedure.

Specifically, in preparing a frequency coordination report, the consultant must coordinate proposed frequency usage with existing and prospective carriers (with applications on file) whose facilities could affect or be affected by a new proposal in terms of frequency interference or "restricted ultimate system capacity." The consultant must identify in its frequency coordination study all area carriers contacted in coordinating its proposed operations.

Following the coordination study, the applicant (through its consultant) must provide the area carriers with the telephone number of its test facility in the unlikely event that the tests should create interference. If there is interference, the applicant must terminate immediately its pattern measurement tests or other operations.

Existing carriers are directed by the FCC to respond promptly to an applicant's notification of a frequency coordination study. The FCC's regulations further provide that if a carrier fails to respond to a request for coordination within 30 days of notification, the applicant will be deemed to have made reasonable efforts to coordinate its facilities and may file its application without obtaining a response from that carrier.

The frequency coordination report must be signed by the person responsible for preparing the report. If the report is prepared by a telecommunications consulting firm, the report should be signed by a technical representative of the firm. Given the lead time associated with this project, it is advisable to commence work on this task as soon as possible prior to the date on which the applicant plans to file the application.

Additional Considerations for International Earth Station Applications

Applications for teleport earth stations to provide IBS and/or international television services must provide the same forms and essentially the same type of information required for domestic earth stations. However, an FCC Form 401 (application for construction permit) must also be submitted.

In addition to the requirement for a Title III application, international earth station applicants must also file a Title II application for authority to serve as a common carrier. This application, which is filed pursuant to section 214 of the Communications Act, should also contain information required by section 63.01 of the FCC's rules.

MICROWAVE LINK APPLICATIONS

General Requirements for Construction Permit

Each microwave facility that is to be installed and operated in connection with the teleport must be licensed by the FCC under Title III of the Communications Act. Part 21 of the FCC rules is applicable to microwave applications. The application for a construction permit for a microwave facility must include an FCC Form 435. The application should also include an FCC Form 430 and appropriate exhibits regarding the applicant's legal qualifications. If the applicant has previously filed a Form 430 for other radio facilities (such as an earth station), the Form 430 and related information can be cross-referenced, provided it has not changed since it was filed. If the information has changed, the applicant must submit the current information.

Generally, the application must demonstrate the applicant's legal, financial, and technical qualifications, disclose the real party in interest, contain a statement regarding whether or not the action is a major environmental action, be appropriately certified and signed by the applicant and by the technical expert, and mention how the facilities will be managed and controlled. The applicable technical standards are contained in part 21 of the FCC's rules.

Other Key Elements of the Application

The required site, cost, and technical information for the application must be provided in connection with the FCC Form 435 and generally includes the following items:

Site availability for station location—owned, and if not, leased, or option to buy or lease—unconditional access.

If new or modified antenna, support structure, antenna characteristics, vertical profile.

FAA notification (state if needed).

Operation and maintenance procedures.

Frequency polarization.

Transmitter output power.

Type of transmitter emission.

Frequency tolerance (types of emissions, bandwidth, emission limitations).

Modulation requirements.

USGS map.

The supplementary showings specific to microwave applications are as follows:

Nature and type of services to be provided.

Cities or communities to be served (where multiple cities to be served—specify by diagram or other appropriate means the circuit cross section between service points).

Projected future circuit growth anticipated between service points and source of such projections.

Other specific requirement applicable only in limited instances.

The financial information required generally includes:

Estimated costs of construction and initial expense.

Estimated operating expenses.

Balance sheet.

Credit arrangement/bank letter.

Application for Operating Authority

The applicant may apply for a license to cover the facilities by filing an FCC Form 436 prior to the expiration date of the construction permit. STAs are available for microwave facilities if immediate operating authority is required. In the absence of mutually exclusive applications, processing of microwave applications is generally similar to that for other radio common carrier facilities.

■   **CONCLUSION**   ■

Because regulatory requirements pertaining to teleport facilities are numerous and complex, it is essential that the teleport developer be advised of all the applicable regulations early in the teleport planning process. The operator can then carefully integrate into that process the resources and time required to comply with those requirements. Failure to do so may result in legal and economic problems and pose serious obstacles to timely implementation of the teleport project.

■ APPENDIX ■

Before the
FEDERAL COMMUNICATIONS COMMISSION
Washington, D.C. 20554

In re Application of	)	
GULF TELEPORT, INC.	)	
for Authority to Construct	)	
and Operate a 14/12 GHz	)	File No.
Transmit-Receive Earth Station	)	
in the Domestic Fixed Satellite	)	
Service in the Houston, Texas,	)	
Metropolitan Area	)	

APPLICATION OF GULF TELEPORT, INC.
FOR CONSTRUCTION PERMIT AND LICENSE
FOR TRANSMIT-RECEIVE EARTH STATION
IN THE DOMESTIC FIXED SATELLITE SERVICE

Andrew D. Lipman
Emilia L. Govan

PEPPER, HAMILTON & SCHEETZ
1777 F Street, N.W.
Washington, D.C. 20006
(202) 842–8100

Counsel for Gulf Teleport, Inc.

February 11, 1985

Before the
FEDERAL COMMUNICATIONS COMMISSION
Washington, D.C. 20554

In re Application of	)	
GULF TELEPORT, INC.	)	
for Authority to Construct	)	
and Operate a 14/12 GHz	)	File No.
Transmit-Receive Earth Station	)	
in the Domestic Fixed Satellite	)	
Service in the Houston, Texas,	)	
Metropolitan Area	)	

To: Common Carrier Bureau
Satellite Radio Branch

APPLICATION

Pursuant to Sections 304, 308, 309, and 319 of the Communications Act of 1934, as amended, 47 U.S.C. §§ 304, 308, 309, and 319 (1982), Gulf Teleport, Inc. ("Gulf Teleport"), by its attorneys, hereby applies for authority to construct and operate a Ku-band (14/12 GHz) transmit-receive earth station in the Domestic Fixed Satellite Service at Algoa (Brazoria County), Texas, a community in the Houston metropolitan area. As demonstrated in this application, Gulf Teleport is legally, technically, and financially qualified to construct and operate this station. Construction of this transmit-receive earth station is a fundamental step in the development of the Gulf Teleport, a multiuser communications facility serving the Houston area and the Texas and Louisiana Gulf Coast region.

■   TABLE OF CONTENTS   ■

THE GULF TELEPORT CONCEPT

Gulf Teleport was organized in September 1983 to design and construct a satellite teleport in the Houston area. The Gulf Teleport site is located approximately 15 miles from downtown Houston and will serve metropolitan Houston as well as the entire Texas/Louisiana Gulf Coast area. The facility will consist of numerous satellite transmit and receive stations providing cost-effective audio, video, and data communication services and will offer Gulf Coast subscribers a number of novel and innovative telecommunications services. In addition to transmit-receive earth stations, Gulf Teleport will utilize terrestrial communications transmitting and receiving equipment, together with various support facilities, computers, and switching and connecting equipment. Moreover, Gulf Teleport will encourage communications-intensive users to locate their back-office and other facilities on the teleport grounds or in an adjacent soon-to-be constructed office park complex.

This application is the second in a series of applications being filed by Gulf Teleport and others for authority to construct and operate earth stations at the Gulf Teleport site. On September 27, 1984, Gulf Teleport filed the first of these applications. That application requested approval for the construction and operation of a C-band (6/4 GHz) transmit-receive earth station at the same location (hereinafter referred to as the "C-band application"). The C-band application was placed on public notice on October 10, 1984 (File No. 002422–DSE–P/L–84).

Gulf Teleport plans to connect major user sites in the Houston area directly to the teleport. The teleport will also provide microwave trunk lines to key commercial and industrial complexes along the entire Gulf Coast, thereby permitting access to its system by users in Beaumont, Port Arthur, Orange, Victoria, and Corpus Christi, Texas, and Lake Charles, Louisiana. An application for authority to construct and operate these microwave facilities will be submitted to the commission at the appropriate time. Gulf Teleport may also use fiber optic transmission facilities to connect the teleport to various sites in the Gulf region.

Gulf Teleport will also offer access to international points. In this regard, Gulf Teleport notes that COMSAT has announced its intent to locate an international earth station at the Gulf Teleport complex, which will provide IBSS (INTELSAT Business Satellite Services) to area users. COMSAT's application for the IBSS earth station was filed with the commission on August 10, 1984, and placed on public notice on August 22, 1984 (File No. CSG–84–036 P/L). It is expected that COMSAT's earth station will be operational later this year. As a result of these and other arrangements, Gulf Teleport plans to serve as both a domestic and international communications gateway for the entire Gulf region.

The earth station applied for in this application—to be operated by Gulf Teleport—is an essential step in bringing the Gulf Teleport concept to fruition. It is anticipated that approval of this application will encourage others to locate their satellite and related communications facilities at the Gulf Teleport complex.[1]

APPLICANT QUALIFICATIONS

Gulf Teleport hereby applies for simultaneous grant of a construction permit and a license to operate the proposed earth station facility. As demonstrated in this application, Gulf Teleport has the requisite legal, financial, and technical qualifications to hold the permit and license for which it is applying. Gulf Teleport intends to commence construction of the earth station immediately after issuance of the construction permit and to complete construction of the facility within six months of the grant of the permit.

Legal Qualifications

Gulf Teleport is legally qualified to construct and operate a transmit-receive earth station, as set forth in FCC Form 430, a copy of which is on file with the FCC as Exhibit A to Gulf Teleport's C-band application. As seen from the biographies of the officers and directors submitted in connection with the Form 430, Gulf Teleport will have the benefit of highly qualified management personnel with extensive experience in telecommunications and related fields.

Financial Qualifications

As shown in Exhibit I to the application, Gulf Teleport is fully capable of meeting the projected costs of constructing and operating the earth station and therefore has the requisite financial qualifications to hold the construction permit and license applied for herein.

Technical Qualifications

As discussed below, the technical characteristics of the proposed earth station are in full compliance with applicable laws and regulations. Further, the Gulf Teleport personnel are technically qualified to construct and operate the facility.

SITE LOCATION AND AVAILABILITY

The proposed earth station facility applied for herein would be located at the Gulf Teleport site on County Road #82 in Algoa, Brazoria County, Texas. The 20-acre Gulf Teleport site is in a rural farming area approximately 15 miles south of the city of Houston and is surrounded by County Road #82 to the west and farmland to the north, east, and south. The geographic coordinates of the earth station site are as follows:

North Latitude	29°	29'	03"
West Longitude	95°	23'	02"

The earth station antenna will have a diameter of approximately 9.2 meters. An equipment shelter occupying an area of approximately 200 square feet will be located adjacent to the antenna. Ground elevation for the proposed site is 55 feet above mean sea level. A 7.5 minute topographic map of the earth station site location is included as Exhibit II.

The Teleport site is currently undeveloped. A sketch of the Gulf Teleport satellite communications tract is included as Exhibit III. The earth station applied for herein will be one of a number to be located at the Teleport site. A total of eight locations at the Teleport are planned as antenna sites, of which four will be constructed, owned, and operated by Gulf Teleport. (Gulf Teleport has previously filed one application and plans to file subsequent applications for these sites at the appropriate time.) The other sites will be leased to other entities for construction and operation of earth stations. As discussed above, one site has already been assigned to COMSAT, whose application for a transmit-receive international satellite earth station is presently on file with the commission.

The proposed site for the earth station is currently owned by Milestone Properties, Inc., Trustee, of Houston. Gulf Teleport and Milestone Properties, Inc. have entered into an Option Contract whereby Gulf Teleport has an option, which expires in June 1985, to purchase this property for use as a teleport facility. Gulf Teleport is not aware of any zoning or municipal regulations that would impair the anticipated use of this land.

FREQUENCY COORDINATION AND INTERFERENCE ANALYSIS

A Frequency Coordination and Interference Analysis Report for the proposed Gulf Teleport earth station location has been prepared by Comsearch, Inc. and is submitted as Exhibit IV to this application. Gulf Teleport will soon complete frequency coordination for each of the other sites in the teleport. As teleport operator, Gulf Teleport will take steps to ensure that other antenna slips within the teleport site are adequately protected from interference from other services or users.

TECHNICAL CHARACTERISTICS

Overall Technical Description

A general, overall technical description of the proposed earth station facilities and operations is contained in Exhibit V to this application. Included in Exhibit V is information about the major structures and equipment composing the station (including a functional block diagram). Exhibit V also contains the estimated capital and operating costs of the proposed earth station.

On-site construction of the earth station should begin immediately following issuance of the construction permit, and it is expected that construction will be completed within six months of the grant of the permit.

Technical Details

Also contained in Exhibit V is a description of the particulars of operations of the earth station, the transmitting equipment, and the antenna facilities.

Points of Communication

The space segment to be accessed is the domestic satellite arc from 55 to 143 degrees west longitude. Access will be arranged with appropriate satellite carriers.

Communications Capacity

The station's communications capacity will consist of: three video channels, three high-rate data channels, and/or equivalent voice capacity.

ENVIRONMENTAL IMPACT

Commission grant of this license would not be a "major" action under the terms of Section 1.1305 of the Commission's Rules, 47 C.F.R. §1.1305*(a)* & *(b)* (1983), since the proposed satellite earth station antenna has a diameter of approximately 9.2 meters and is located in an "antenna farm."

RADIATION HAZARD ANALYSIS

A statement concerning the radiation characteristics of the proposed earth station is attached as Exhibit VI.

APPLICATION FOR RADIO STATION LICENSE

An application for radio station license (FCC Form 403) is attached as Exhibit VII.

PUBLIC INTEREST SHOWING

As demonstrated in this application, the public interest will be substantially furthered by the grant of a permit and license for construction and operation of this Ku-band transmit-receive earth station by Gulf Teleport. Among other things, the proposed facility will significantly contribute to satisfying the growing and unsatisfied needs of Houston and other Gulf area communities for access to satellite communications. Given the high frequency congestion in this area, this need for spectrally efficient communications is particularly acute.

The earth station applied for herein will constitute an important component of the Gulf Teleport. It is anticipated that favorable processing of this application will further encourage other carriers to collocate their satellite earth stations and other communications facilities at this site. This will further the teleport's goal to accommodate many of the telecommunications needs of the Houston and Gulf areas while providing the cost and spectrum efficiencies of collocating multiple earth stations at a single site.

By providing cost-efficient voice, data, and video communications among users in the Houston region and between them and other users throughout the nation and in the international business community, the Gulf Teleport will provide a valuable communications gateway. This will encourage the development of high technology regional commercial real estate development and attract clean, growth-oriented high technology industries to the Houston area.

For all these reasons, the proposed Ku-band earth station, as well as the planned Gulf Teleport, would further the commission's obligation to promote rapid and efficient communications services at reasonable cost, and the commission's policy of furthering competition in the telecommunications marketplace.

CORRESPONDENCE

All correspondence or inquiries relating to this application should be addressed to:

> Andrew D. Lipman, Esquire
> PEPPER, HAMILTON & SCHEETZ
> 1777 F Street, N.W.
> Washington, D.C. 20006
> Phone: (202) 842–8110

with a copy to:

> Olan Jones
> President,
> GULF TELEPORT, INC.
> 7541 South Freeway
> Houston, Texas 77021
> Phone: (713) 748–5530

SECTION 304 WAIVER

Pursuant to Section 304 of the Communications Act of 1934, as amended, 47 U.S.C. §304 (1983), Gulf Teleport hereby waives any claim to the use of any particular frequency or of the ether as against the regulatory power of the United States because of the previous use of the same, whether by license or otherwise.

CONCLUSION

As demonstrated in this application, Gulf Teleport is legally, financially, and technically qualified to construct and operate the proposed earth station, as well as the planned teleport, in the public interest. Gulf Teleport's application for a construction permit and license for a transmit-receive earth station in the Domestic Fixed Satellite Service to be located at the Gulf Teleport site in Algoa, Texas, in the Houston metropolitan area should therefore be granted.

Respectfully submitted,

GULF TELEPORT, INC.

By _Andrew D. Lipman_

Andrew D. Lipman
Emilia L. Govan

PEPPER, HAMILTON & SCHEETZ
1777 F Street, N.W.
Washington, D.C. 20006
(202) 842–8100

Counsel for Gulf Teleport, Inc.

February 11, 1985

EXHIBIT I
Financial Qualifications

Gulf Teleport is financially qualified to construct and operate the proposed earth station. Included in this exhibit is Gulf Teleport's current balance sheet.

A demonstration of Gulf Teleport's additional financial qualifications, including letters from the Perpetual American Bank and the Commerce Bank–Del Oro, is contained in Exhibit B to Gulf Teleport's C-band application and is incorporated herein by reference.

GULF TELEPORT, INC.
(A Development Stage Enterprise)
Balance Sheet
October 31, 1984

Assets

Current assets:	
Cash	$ 40,461
Certificates of deposit	308,965
	349,426
Furniture and fixtures, net	493
Acquisition and development costs	355,236
Other assets, net	603
	$705,758

Liabilities and Stockholders' Equity

Current liabilities:	
Accounts payable	$ 84,842
Commitments and contingencies (notes 3, 5, and 6)	
Stockholders' equity (notes 4, 5, and 6):	
Common stock $.001 par value; authorized 25,000,000 shares; issued and outstanding, 9,966,667 shares....................	9,967
Capital in excess of par value	776,438
Deficit accumulated during developmental stage.................	(165,489)
	620,916
	$705,758

EXHIBIT II
7.5-Minute Topographic Map

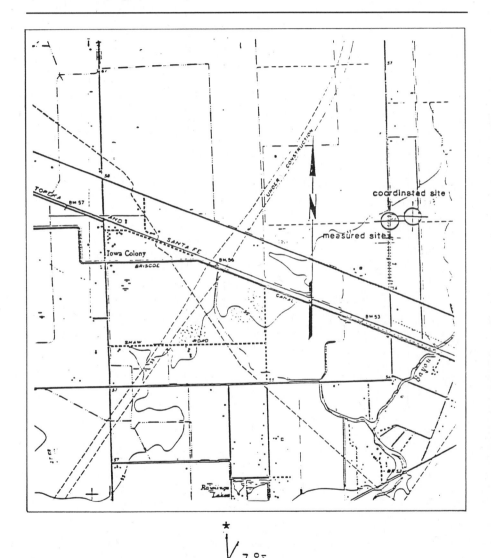

EXHIBIT III
Gulf Teleport Tract

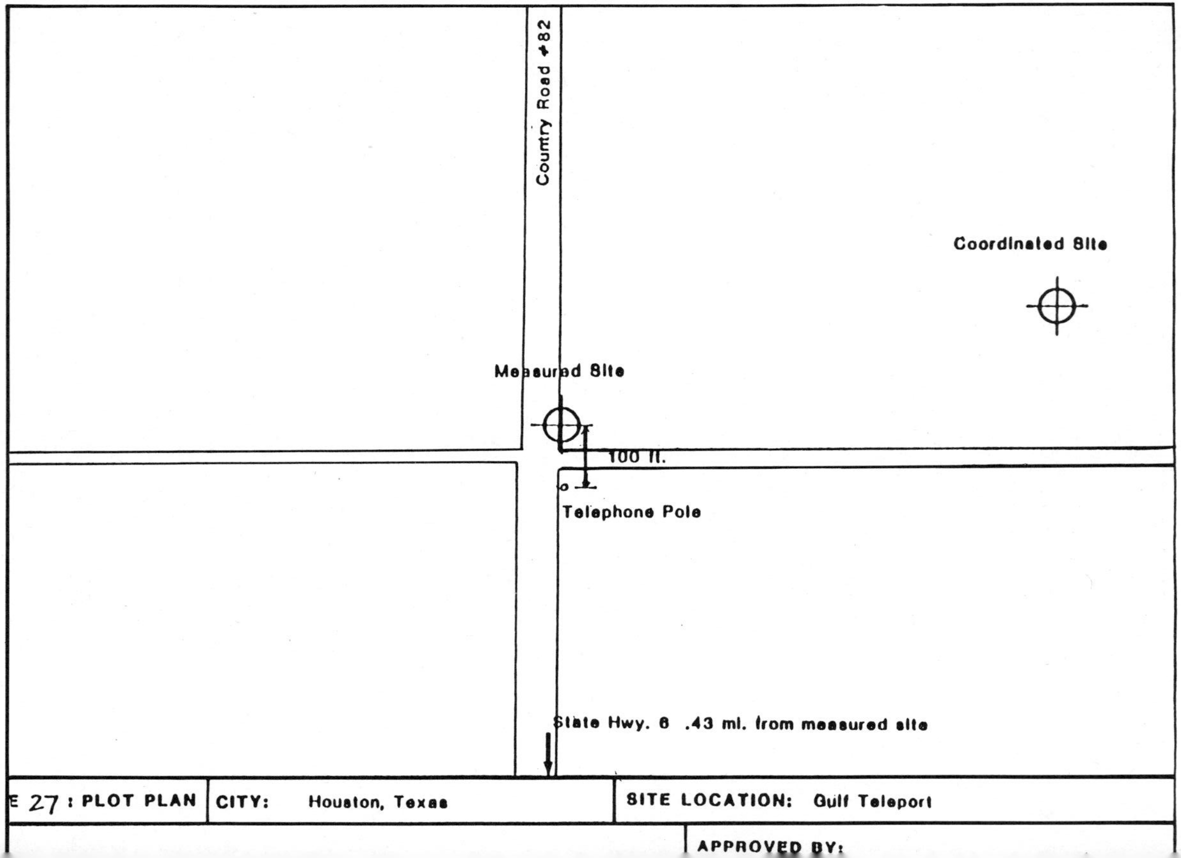

Coordinated Site

Country Road #82

Measured Site

100 ft.

Telephone Pole

State Hwy. 8   .43 ml. from measured site

27 : PLOT PLAN | CITY: Houston, Texas | SITE LOCATION: Gulf Teleport

APPROVED BY:

EXHIBIT IV
Frequency Coordination Report

Prepared for Gulf Teleport, Inc., Houston, Texas, Satellite Earth Station.

Prepared by Comsearch, Inc., 11503 Sunrise Valley Drive, Reston, Virginia 22091, January 28, 1985.

Contents

1. Supplemental showing, re: Part 25.203(C)
2. Earth station coordination data
3. Certification

Supplemental Showing, Part 25.203(C)
Gulf Teleport, Inc.
Houston, Texas

Pursuant to part 25.203(C) of the FCC rules and regulations the above referenced satellite earth station was coordinated with the existing licensees and applicants whose facilities could be affected. Coordination data was forwarded on January 28, 1985.

The following carriers or their designated coordination agents were notified:

Bell Communications Research Inc.
The Western Union Telegraph Company
AT&T Communications
AT&T Communications Southern Region
United Satelco Transmission, Inc.
AT&T Communications Midwestern Region
Comsat General Corporation

There are no unresolved interference objections with the stations contained in these applications.
Respectfully Submitted,
Comsearch, Inc.

Kenneth D. Ryan

Kenneth G. Ryan
Frequency Coordinator

EXHIBIT IV *(continued)*

Earth Station Coordination Data

This section presents the data pertinent to frequency coordination of the proposed earth station, which was circularized to all common carriers within its coordination contours.

Applicant:	Gulf Teleport, Inc.
Earth Station Name:	Houston, TX
Latitude (DMS):	29 29 3
Longitude (DMS):	95 23 2
Site ground elevation (ft. AMSL):	55.
Antenna centerline (AGL):	16.
Antenna type:	Harris 9.0 Meter
12 GHz gain (DBI)/diameter (meters):	59.2/ 9.0
3 DB/15 DB half beamwidth (deg.):	0.2/ 0.1
14 GHz gain (DBI)/diameter (meters):	60.4/ 9.0
3 DB/15 DB half beamwidth (deg.):	0.2/ 0.1
Operating mode:	T/R
Receive band (GHz):	11.7 − 12.2
Transmit band (GHz):	14.0 − 14.5
Emission designator:	72000F9 & 72000F9Y
Modulation:	Analog & digital
Max. available RF power (DBW/4kHz):	−3.0
(DBW/MHz):	21.0
Max. EIRP (DBW/4 kHz):	57.4
(DBW/MHz):	81.4
Max. permissible interference power	
12 GHz, 20% (DBW):	−156.0
12 GHz, 0.0150% (DBW):	−144.0
14 GHz, 20% (DBW/4kHz):	−154.0
14 GHz, 0.0025% (DBW/4kHz):	−131.0
Satellite arc (Min./max.):	65.0/ 143.0
Azimuth (min./max.):	127.8/ 245.8
Elevation:	40.7/ 28.3
Radio Climate:	A
Rain zone:	2
Maximum great circle coordination distance (km)	
12 GHz:	228.0
14 GHz:	177.2
Precipitation scatter contour radius (km)	
12 GHz:	481.1
14 GHz:	100.0

Note: Horizon is less than 0.2 degrees at all azimuths.

EXHIBIT IV (continued)

Houston, Texas

9 29 3   95 23 2

01/28/85

Azimuth Degrees	Horizon Angle Degrees	Disc. Angle Degrees	Horizon Gain 12 GHz DBI	12 GHz Coord. Distance km	Horizon Gain 14 GHz DBI	14 GHz Coord. Distance km
0.	0.0	111.2	−10.0	203.4	−10.0	158.9
5.	0.0	114.2	−10.0	203.4	−10.0	158.9
10.	0.0	110.7	−10.0	203.4	−10.0	158.9
15.	0.0	107.1	−10.0	203.4	−10.0	158.9
20.	0.0	103.4	−10.0	203.4	−10.0	158.9
25.	0.0	99.7	−10.0	203.4	−10.0	158.9
30.	0.0	95.9	−10.0	203.4	−10.0	158.9
35.	0.0	92.1	−10.0	203.4	−10.0	158.9
40.	0.0	88.3	−10.0	203.4	−10.0	158.9
45.	0.0	84.6	−10.0	203.4	−10.0	158.9
50.	0.0	80.8	−10.0	203.4	−10.0	158.9
55.	0.0	77.1	−10.0	203.4	−10.0	158.9
60.	0.0	73.4	−10.0	203.4	−10.0	158.9
65.	0.0	69.7	−10.0	203.4	−10.0	158.9
70.	0.0	66.2	−10.0	203.4	−10.0	158.9
75.	0.0	62.7	−10.0	203.4	−10.0	158.9
80.	0.0	59.4	−10.0	203.4	−10.0	158.9
85.	0.0	56.2	−10.0	203.4	−10.0	158.9
90.	0.0	53.2	−10.0	203.4	−10.0	158.9
95.	0.0	50.4	−10.0	203.4	−10.0	158.9
100.	0.0	47.9	−10.0	203.4	−10.0	158.9
105.	0.0	45.7	−9.5	205.4	−9.5	160.4
110.	0.0	43.8	−9.0	207.3	−9.0	161.8
115.	0.0	42.4	−8.7	208.9	−8.7	162.9

EXHIBIT IV (continued)

Azimuth Degrees	Horizon Angle Degrees	Disc. Angle Degrees	Horizon Gain 12 GHz DBI	12 GHz Coord. Distance km	Horizon Gain 14 GHz DBI	14 GHz Coord. Distance km
120.	0.0	41.3	−8.4	210.0	−8.4	163.8
125.	0.0	40.8	−8.3	210.6	−8.3	164.2
130.	0.0	40.8	−8.3	210.6	−8.3	164.2
135.	0.0	41.3	−8.4	210.1	−8.4	163.8
140.	0.0	42.2	−8.6	209.0	−8.6	163.1
145.	0.0	43.6	−9.0	207.5	−9.0	162.0
150.	0.0	45.4	−9.4	205.7	−9.4	160.6
155.	0.0	47.6	−9.9	203.6	−9.9	159.0
160.	0.0	50.1	−10.0	203.4	−10.0	158.9
165.	0.0	52.4	−10.0	203.4	−10.0	158.9
170.	0.0	54.1	−10.0	203.4	−10.0	158.9
175.	0.0	55.2	−10.0	203.4	−10.0	158.9
180.	0.0	55.6	−10.0	203.4	−10.0	158.9
185.	0.0	55.2	−10.0	203.4	−10.0	158.9
190.	0.0	54.1	−10.0	203.4	−10.0	158.9
195.	0.0	52.4	−10.0	203.4	−10.0	158.9
200.	0.0	50.1	−10.0	203.4	−10.0	158.9
205.	0.0	47.4	−9.9	203.9	−9.9	159.2
210.	0.0	44.2	−9.1	206.9	−9.1	161.5
215.	0.0	40.9	−8.3	210.5	−8.3	164.2
220.	0.0	37.5	−7.4	214.4	−7.4	167.1
225.	0.0	34.6	−6.5	218.3	−6.5	170.0
230.	0.0	32.1	−5.7	221.9	−5.7	172.7
235.	0.0	30.1	−5.0	224.9	−5.0	175.0
240.	0.0	28.8	−4.5	227.1	−4.5	176.6
245.	0.0	28.3	−4.3	228.0	−4.3	177.2

250.	0.0	28.6	−4.4	227.6	−4.4	176.9
255.	0.0	29.6	−4.8	225.8	−4.8	175.6
260.	0.0	31.4	−5.4	223.0	−5.4	173.5
265.	0.0	33.7	−6.2	219.5	−6.2	170.9
270.	0.0	36.5	−7.1	215.7	−7.1	168.0
275.	0.0	39.7	−8.0	211.8	−8.0	165.1
280.	0.0	43.2	−8.9	207.9	−8.9	162.2
285.	0.0	47.0	−9.8	204.2	−9.8	159.5
290.	0.0	50.8	−10.0	203.4	−10.0	158.9
295.	0.0	54.9	−10.0	203.4	−10.0	158.9
300.	0.0	59.0	−10.0	203.4	−10.0	158.9
305.	0.0	63.2	−10.0	203.4	−10.0	158.9
310.	0.0	67.5	−10.0	203.4	−10.0	158.9
315.	0.0	71.8	−10.0	203.4	−10.0	158.9
320.	0.0	76.1	−10.0	203.4	−10.0	158.9
325.	0.0	80.5	−10.0	203.4	−10.0	158.9
330.	0.0	84.9	−10.0	203.4	−10.0	158.9
335.	0.0	89.3	−10.0	203.4	−10.0	158.9
340.	0.0	93.7	−10.0	203.4	−10.0	158.9
345.	0.0	98.1	−10.0	203.4	−10.0	158.9
350.	0.0	102.5	−10.0	203.4	−10.0	158.9
355.	0.0	106.8	−10.0	203.4	−10.0	158.9

EXHIBIT IV *(continued)*

Houston, Texas
Gulf Teleport, Inc.

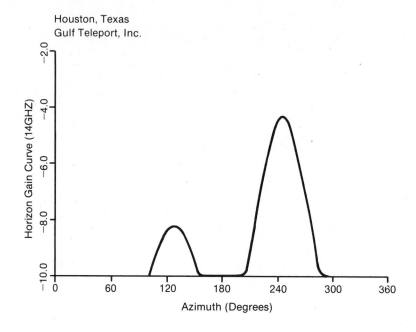

Houston, Texas
Gulf Teleport, Inc.

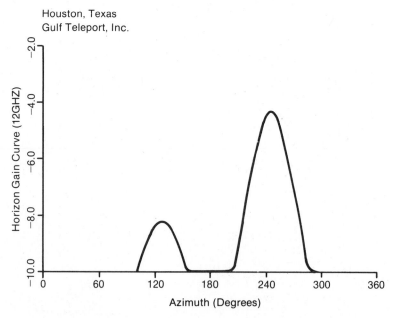

EXHIBIT IV *(continued)*

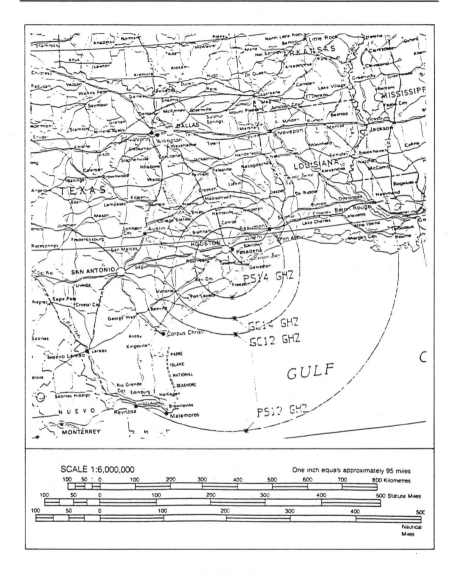

Certification

I hereby certify that I am the technically qualified person responsible for the preparation of the frequency coordination data contained in this application, that I am familiar with parts 21 and 25 of the FCC Rules and Regulations, that I have either prepared or reviewed the frequency coordination

---

EXHIBIT IV *(concluded)*

---

data submitted with this application, and that it is complete and correct to the best of my knowledge and belief.

By: *Kenneth D Ryan*

Kenneth G. Ryan
Frequency Coordinator
Comsearch, Inc.
11503 Sunrise Valley Drive
Reston, Virginia 22091

Dated: *January 28, 1985*

EXHIBIT V
Technical Information and Estimated Costs

FUNCTIONAL DESCRIPTION

The Gulf Teleport Ku-band earth station will provide facilities for carrying commercial communications via any U.S. domestic Ku-band satellite system.

A cost-effective arrangement of equipment will be provided as shown in Figure 1 (System Block Diagram). This will include the following:

a. A nominal 9.2-meter cassegrain type, limited-motion antenna equipped with step-track capability.
b. Four 2-kW Klystron amplifiers in a one-for-two redundant configuration.
c. Two low-noise amplifiers (190°K nominal noise temperature) in a one-for-one redundant configuration.
d. Broadband GCE subsystem consisting of one-for-two redundant transmit and receive configuration.
e. Monitor and control facilities with remote monitor and control capabilities for unattended operation.

Antenna System

A cassegrain type antenna having a nominal diameter of 9.2 meters will be provided. The mount will be a standard elevation over azimuth type.

EXHIBIT V *(continued)*

The antenna will be capable of at least ±55° of angular travel in azimuth and will have a range in elevation extending from 5° to 90°. Polarization will be linear, with transmit and receive bands having opposite sense. The antenna will comply with a cross-polarization isolation specification of at least 35 dB (on-axis). This type and size of antenna is available from several manufacturers.

The station G/T will be at least 34.2 dB/°K at 12.0 GHz at all operating elevation angles. The antenna will have a nominal gain of 60.2 dBi at 14.0 GHz and 59.1 dBi at 11.7 GHz. The system noise temperature will generally be less than 310°K. The antenna will comply with the FCC's recently revised sidelobe regulations. See 47 C.F.R. Section 25.209(a).

A step track system will be provided.

High-Power Amplifiers

To meet the projected uplink EIRP requirements, 2-kW Klystron amplifiers will be used. The power amplifiers have been sized to include traffic growth capability. Each 2-kW high-power amplifier can transmit carriers to other typical teleport and privately owned earth stations with sufficient operating margin.

Low-Noise Amplifiers

The nominal 9.2-meter antenna requires the employment of reliable low-noise amplifiers with a nominal noise temperature of 190°K or less to meet the G/T requirement. The amplifiers will be arranged in a one-for-one redundant configuration as shown in Figure 1.

Broadband GCE Subsystem

The receive GCE subsystem will be arranged for one-for-three redundancy. Each redundant chain will consist of a dual-conversion downconverter and the required interface equipment.

The transmit GCE subsystem will be arranged for one-for-two redundancy. Each redundant chain will consist of a dual-conversion upconverter and the required interface equipment.

Baseband Equipment

Transmission and reception of signals will be made using FM or quadriphase shift keying (QPSK) modulation. Configuration of baseband equipment will depend on specific services ordered.

FIGURE 1
System Block Diagram

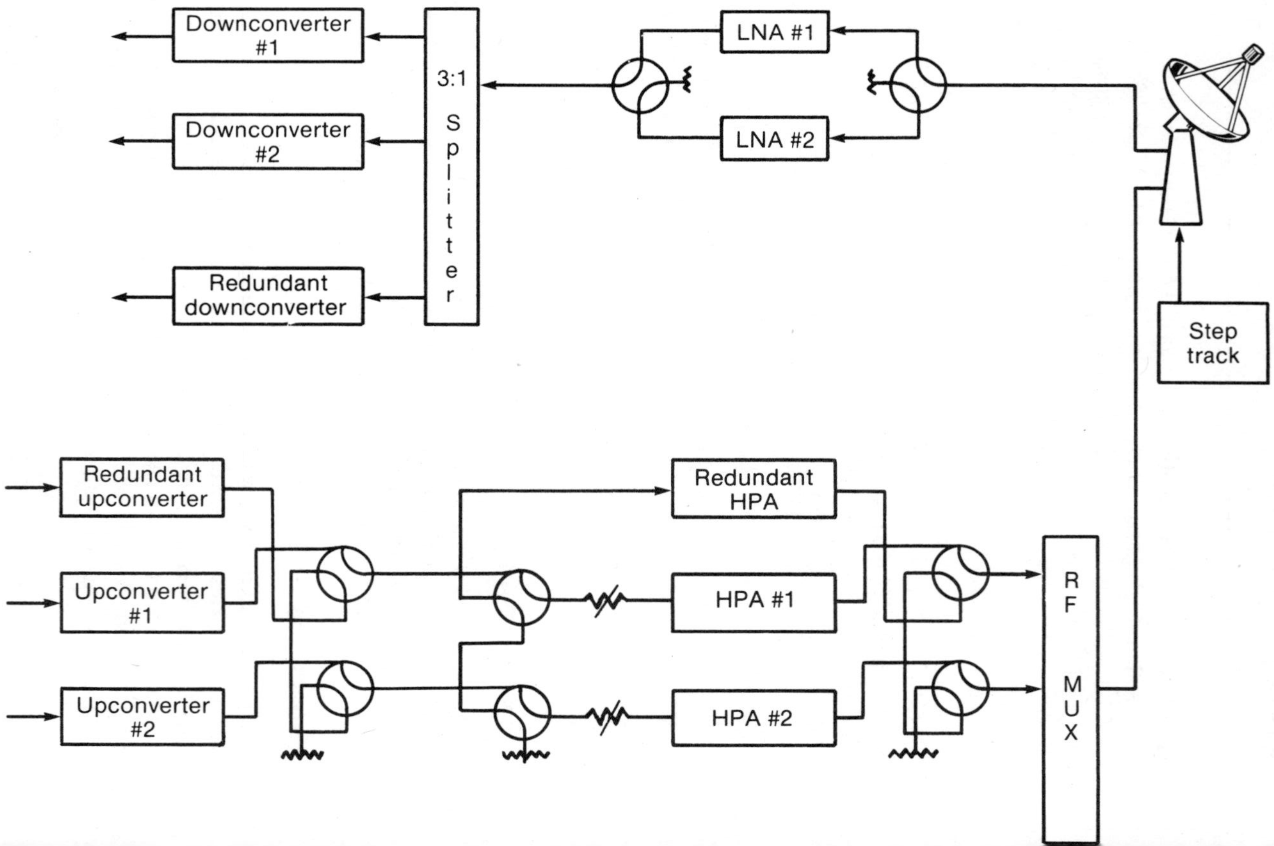

EXHIBIT V *(continued)*

Power System

Prime power will be supplied by the local utility company, with backup power generated on-site. An uninterruptible power system will provide continuity of service in the event of an interruption of commercial power.

Monitor and Control System

The monitor and control system will consist of a centralized monitor and control mimic panel providing alarm and status information of all the earth station subsystems.

In addition, a processor located at the station will automatically report all major faults to a manned teleport facility, which will have the capability to turn off RF power at the station.

Communications Equipment Shelter

All communications equipment except the LNAs will be contained in a communications equipment shelter located adjacent to the antenna. The equipment shelter will initially have approximately 200 square feet of area with capability for future expansion as necessary.

Earth Station Characteristics

I. General
    A. Station name: to be selected.
    B. Call sign: to be determined.
    C. Nature of service: communications satellite.
    D. Class of station: fixed earth station.
    E. Location of station: Houston, Texas.
    F. Geographical coordinates of station site: 29° 29′ 03″ N. latitude, 95° 23′ 02″ W. longitude.

II. Particulars of operation

Frequency MHz	Emission	Normal EIRP	Maximum EIRP Density
14,000–14,500	72,000 F9 to 72,000 F9Y	67.2 dBW to 87.2 dBW	46.9 dBW/4 kHz

    A. Range of center of main lobe of radiation with respect to true north: 95° to 130°.

EXHIBIT V *(continued)*

    B. Range of elevation angles: 5° to 90°.
    C. Range of satellite longitudes: 55° to 143° west longitude.
III. Transmitting requirements

Type	Output Power	Frequency Tolerance
[2] Klystron	2,000 Watts	0.001%

IV. Antenna Facilities
    A. Type: SATCOM Technologies Model 920K 9.2-meter nominal diameter parabolic antenna (or equivalent).
    B. Frequency range: 11,700 to 12,200 MHz (transmit), 14,000 to 14,500 MHz (receive)
    C. Antenna gain: 59.1 dBi at 11,700 MHz, 60.2 dBi at 14,000 MHz.
    D. Beamwidth: Half power beamwidth of 0.18° at 11,700 MHz, 0.15° at 14,000 MHz.
    E. Site elevation: 55 feet AMSL.
    F. Antenna center height: 14 feet AGL, 69 feet AMSL.
    G. Maximum antenna height: 33 feet AGL, 88 feet AMSL.
    H. Polarization: Linear, horizontal and vertical.
    I. Receiving system noise temperature: 309° at 20° elevation.
    J. Gain-to-noise temperature: 34.2 dB/°K or better at 12.0 GHz at 20° elevation.

Predicted Performance of 9.2-Meter Parabolic Antenna

	Frequency Band	
Item	14 GHz	12 GHz
Frequency range:	14,000–14,500 MHz	11,700–12,200 MHz
Polarization:	Linear, horizontal, and vertical	Linear, horizontal, and vertical
Gain:	60.2 dBi	59.1 dBi
VSWR:	1.3:1	1.3:1
Side lobes:	Meets FCC rules part 25.209(a)	Meets FCC rules part 25.209(a)
Cross-polarization Isolation:	35 dB (minimum on-axis)	35 dB (minimum on-axis)

EXHIBIT V *(continued)*

SYSTEM OPERATIONS

The services to be provided to the business community of the Gulf Teleport earth station will include data and voice communications in a digital form.

Preliminary market studies performed by the Gulf Teleport indicate that in the near future there will be demand for the transmission and reception of the following types of signals:

1. Low-speed data.
2. High-speed data.
3. Voice.
4. Full-motion video.
5. Limited-motion (teleconferencing) video.

The above signals will be received at the Gulf Teleport earth station site via a terrestrial microwave facility.

GULF TELEPORT, INC.
Estimated Capital Costs

Transmitting and receiving	$ 300,000
Antenna and waveguide	350,000
Power plant, control, and common equipment	125,000
Buildings and towers	75,000
Spare parts and miscellaneous	150,000
Total	$1,000,000

GULF TELEPORT, INC.
Estimated Annual Operating Expense

Electrical power	$13,000
Travel	5,000
Telephone	5,000
Salaries	40,000
Payroll taxes	4,000
Office expenses	1,000
Consulting fees	5,000
Insurance	3,000
Total	$76,000

TECHNICAL CERTIFICATION

I hereby certify that I am the technically qualified person responsible for preparation of the engineering information contained in this application; that I am familiar with parts 21 and 25 of the Commission's Rules; that I have

---

EXHIBIT V *(concluded)*

---

either prepared or reviewed the engineering information submitted in this application; and that it is complete and accurate to the best of my knowledge.

By _____

Name _____Bartus H. Batson_____

Title _____President_____

Company __Communications & Data Systems Assoc.

Date _____December 15, 1984_____

EXHIBIT VI
Radiation Hazard Analysis

RADIATION HAZARD ANALYSIS

9.2-Meter Parabolic Earth Station Antenna

In compliance with the Commission's Memorandum, Opinion and Order, Section V, Paragraph 95 (Docket No. 16495), adopted December 21, 1972, studies were conducted to predict radiation levels around the proposed earth station. The Occupational Safety and Health Act of 1970 defines the maximum level of nonionizing radiation to which employees may be exposed as a power denisty of 10 milliwats per square centimeter (10mW/cm²) averaged over any six-minute period, as derived from Standard C95.1 of the American National Standards Institute (ANSI). Following are calculations that provide the radiation levels for the proposed earth station antenna system.

Far-Field Region

Free-space power density is maximum on-axis and may be calculated from equation 1.

$$PFD_{ff} = \frac{GP_t}{4 \pi R^2} \tag{1}$$

EXHIBIT VI *(continued)*

---

where

PFD_{ff} = The power flux density on-axis in the far field.
G = The on-axis gain of the antenna, which for a 9.2-meter antenna at 14 GHz will be 60.2 dBi.
P_t = The maximum transmitted power, which will be 501 watts (27.0 dBW).
R = The distance to the far field region and is found from equation 2.

$$R = \frac{2D^2}{\lambda} \qquad (2)$$

where

D = Antenna diameter = 9.2 meters.
λ = Wavelength at 14.00 GHz = 0.021 meters = 2.1 cm.

From equation 2 it is found that the distance to the far field is 8,061.0 meters, and, PFD_{ff} is found from equation 1 as follows:

$$
\begin{aligned}
PFD_{ff} &= 60.2 \text{ dB} + 27.0 \text{ dBW} - 89.1 \text{ dB/m}^2 \\
&= -1.9 \text{ dBW/m}^2 \\
&= -41.9 \text{ dBW/cm}^2 \\
&= -11.9 \text{ dBm/cm}^2 \\
&= 0.065 \text{ mW/cm}^2
\end{aligned}
$$

Near-Field Region

The geometrical limits of the radiated power in the near field approximate a cylindrical volume with a diameter equal to that of the antenna. In the near field, the power density is neither uniform nor does its value vary uniformly with distance from the antenna. For the purpose of considering radiation hazard, therefore, it is assumed that the on-axis flux density is at its maximum value throughout the length of this region. The length of this region, that is, the distance from the antenna to the end of the near field, is given by the following.

$$L_{nf} = \frac{A}{2\lambda} = \frac{\pi D^2}{8\lambda} \qquad (3)$$

where

L_{nf} = Length to the end of the near field.
A = Cross-sectional area of the antenna.
D = Diameter of antenna = 9.2 m = 920 cm.

EXHIBIT VI *(concluded)*

Substituting,

$$L_{nf} = \frac{\pi(920)^2}{8 \times 2.1} = 158{,}280 \text{ cm} = 1582 \text{ m}$$

The maximum power flux density in the near field is given by:

$$PDF_{nf} = \frac{16P_t}{\pi D^2} \tag{4}$$

where P_t is the maximum power transmitted by the antenna (27.0 dBW or 57.0 dBm). From equation (4), we see that:

$$PDF_{nf} = 12.04 \text{ dB} + 57.0 \text{ dBm} - 4.97 \text{ dB} - 59.3 \text{ dBcm}^2$$
$$= 4.77 \text{ dBm/cm}^2$$
$$= 3.0 \text{ mW/cm}^2$$

Conclusion

The study shows that with the operational characteristics of the proposed system, the power density levels do not exceed the 10 mW/cm² OSHA standard, even in the near field of the antenna. This near-field region exists from the antenna to a distance of 1,582 meters, as shown in the above calculations, and the maximum power density level is 3.0 mW/cm².

EXHIBIT VII
Application for Radio Station License (FCC Form 403)

Approved by OMB
3060-0019
Expires 9-30-84

	APPLICANT SHOULD NOT USE THIS BLOCK	
	File Number	Call Sign

Federal Communications Commission
Washington, D. C. 20554

APPLICATION FOR RADIO STATION LICENSE
OR
MODIFICATION THEREOF
UNDER PARTS 22, 23, OR 25

INSTRUCTIONS

A. Submit TWO COPIES of this application direct to the Federal Communications Commission, Washington, D. C. 20554, for radio licenses in the following services (check appropriate box below).

☐ **FCC Rules Part 22** - Domestic Public Radio Services
(Other than Maritime Mobile)

☐ **FCC Rules Part 23** - International Fixed
Public Radiocommunication Services

☒ **FCC Rules Part 25** - Satellite Communications

B. It is recommended that, before submitting application, applicant refer to the appropriate rule part which governs the class of station applied for. Copies of the rules may be purchased from the Superintendent of Documents, Government Printing Office, Washington, D. C. 20402.

C. Use additional sheets only where necessary. Such sheets must be marked as exhibits and referred to in the application.

NOTIFICATION TO INDIVIDUALS UNDER PRIVACY ACT OF 1974
AND THE PAPERWORK REDUCTION ACT OF 1980

The information requested by this form will be used by Federal Communications Commission staff to determine eligibility for issuing authorizations in the use of frequency spectrum and to effect the provisions of regulatory responsibilities rendered the Commission by the Communications Act of 1934, as amended. Information requested by this form will be available to the public. Response to the information requested is required to obtain the requested authorization.

1(a) Name of Applicant:

    Gulf Teleport, Inc.

(b) Mailing Address (number, street, city, state, Zip Code):

    7541 South Freeway
    Houston, Texas   77021

2(a) Class of Station and Call Sign:
    Fixed Earth Station

(b) Nature of Service
    Domestic Fixed Satellite

(c) Purpose of application:

    License to cover construction permit   ☐

    Modification of License   ☐

    Present File No. _____

(d) If for modification of license indicate proposed change:

    Change in frequencies   ☐

    Change in authorized power   ☐

    Change(s) of control point(s)   ☐

    Change in points of communication   ☐

    Change in other particulars   ☐
    (Describe under Remarks on Page 4)

3(a) List the outstanding construction permit(s), if any, which this application covers:

File Number	Date	Call Sign	Manufacturer of Transmitter	Type No.	Serial No.
See Accompanying Application For Construction Permit					

FCC Form 403
December 1983

EXHIBIT VII *(continued)*

(*b*) **If licensed transmitters are being deleted or replaced, show the following with respect to such transmitters:**

Manufacturer	Type No.	Serial No.
N/A		

(*c*) When was the construction specified in 3(*a*) completed? .... N/A ....

(*d*) Is the station now ready for operation? . . . . YES ☐ NO ☒

(*e*) Have all the terms of the construction permit(s) listed in 3(*a*) been met? . . . YES ☐ NO ☒

(*f*) Are all the statements made in the applications for the construction permits or the modifications thereof mentioned in 3(*a*) still true as of the date of this application? . . . . YES ☒ NO ☐

(*g*) If the answer to either or both 3(*d*) and 3(*e*) above is "no", the discrepancies must be shown in appropriate places in this form or listed separately in exhibits and submitted as a part of this form.

Indicate method of submission below:

Numbers of paragraphs containing corrected data ... See accompanying application

Identification of exhibits containing corrected data ... for construction permit, filed ... contemporaneously with this application.

4. Specify in the table *all* particulars of operation exactly as they are desired in the license or modification thereof.

(1) FREQUENCIES AND ANTENNA POLARIZATION	(2) POWER	(3) EMISSION	(4) MODULATING FREQUENCY (Cycles)	(5) TRANSMISSION SPEED (Bauds)	(6) POINTS OF COMMUNICATION
			0-8.5 MHz (video)	Up to 50 MBPS (25 M Baud)	Domestic Satellite Arc
See Exhibit V		See Exhibit V			
					55° to 143° W. Longitude

COLUMN NOTES:

(1) List all frequencies, indicating whether kilohertz or megahertz, and polarization of radiated signal.

(2) Specify whether watts or kilowatts. In the Experimental Radio Services specify effective radiated power and in case of pulse emission peak power.

(3) List all types of emission desired for each frequency. Describe special emission in space below.

Give maximum modulating frequency for each type of emission involved.

(4) Give maximum transmission speed employed in normal operation opposite each type of emission involved. To convert transmission

(5) speed of Continental Morse to bauds, multiply the number of words per minute by 0.8.

(6) Show below the operating agency at each point of communication.

EXHIBIT VII *(continued)*

5. If this application is for authority to operate with *an operator on duty at control point(s) other than the transmitter location -*   N/A
   (a) What will be the location of the control point(s)?

     State _____   County _____

     City or Town _____   Street and No. _____

   (b) What will be the airline distance between transmitter location and the control point(s)? _____

   (c) By what means will the station be monitored while in operation?

     _____

     _____

   (d) Can the transmitter be shut down by the licensed operator at the control point so as to prevent operation from other point(s)? · · · · · · · · YES ☐   NO ☐

   (e) How will unauthorized persons be prevented from having access to the transmitter?

     _____

     _____

6. Proposed location of transmitter:

   (a) *If portable:* ☐   *mobile:* ☐ *(check one, if applicable)* give geographical area of proposed operation:

               N/A

     _____

   (b) If permanently located at a fixed location, give:

     State __Texas_____   County __Brazoria_____

     City or town ____Algoa_____   Street and number __County Road

     N. Latitude: Degrees __29_____, minutes ____29_____, seconds ____3_____

     W. Longitude: Degrees _95_____, minutes ____23_____, seconds ____2_____
           (Give latitude and longitude correct to seconds.)

7. Note any alteration in transmitter(s) or antenna systems not previously reported to the Commission.

         N/A

     _____

8. *(a)* Have there been any changes in the data furnished in the application for construction permit covering ownership, citizenship, station control, business connections, and monopolistic practices? · · · · · · · · · · · · · · YES ☐   NO ☒

   (b) Have such changes been reported to the Commission? If not, such data must be submitted herewith · · · · · · · · · · · · · · · · · · · YES ☐   NO ☒

EXHIBIT VII *(continued)*

9. *(a)* Is station to be open to public correspondence? .... YES ☒ NO ☐

  If so, state hours during which station will be open for such service ....
  .... 24 Hours
  ....

  (b) Will any charge be made for handling public correspondence? .... YES ☐ NO ☐

  If so, state schedules of charges ...Not required as per FCC Docket 79-252....
  The statement of rates required herein does not constitute a filing of schedules of charges required by Section 203 of the Communications Act of 1934, as amended, prior to commencing service.

  (c) State basis of division of charges with other stations ......N/A...

10. If this application is for modification of license, state why the proposed change(s) is (are) deemed necessary and the purpose(s) it will serve.
  ....N/A..
  ....

11. The applicant should be advised that Section **22.13**(f)(2) of the Commission's Rules requires licensees to abide by all State requirements of certification and be in operation within 240 days of the date of the license grant, or the license will automatically expire and must be submitted for cancellation.

  THE APPLICANT hereby waives any claim to the use of any particular frequency or of the ether as against the regulatory power of the United States because of the previous use of the same, whether by license or otherwise, and requests a station license in accordance with this application.

  All the attached exhibits are a material part hereof and are incorporated herein as if set out in full in the application. All the answers on this application are a material part of the application.

CERTIFICATION

  I certify that the statements in this application are true, complete, and correct to the best of my knowledge and belief, and are made in good faith.

  Signed and dated this 5 day of ___February___ , 19 85

  Gulf Teleport, Inc.
  _____
  Name of Applicant *(must correspond with item 1a)*

  Olan C. Jones   /s/
By _____
  Signature *(designate by checkmark below appropriate classification)*

  ┌─────────────────────────────────────┐
  │ WILLFUL FALSE STATEMENTS MADE ON   │   ☐ INDIVIDUAL APPLICANT
  │ THIS FORM ARE PUNISHABLE BY FINE AND │   ☐ MEMBER OF APPLICANT PARTNERSHIP
  │ IMPRISONMENT. U. S. CODE, TITLE 18   │   ☐ OFFICER OF APPLICANT CORPORATION OR OFFICER AND
  │ SECTION 1001.   │   MEMBER OF APPLICANT ASSOCIATION
  └─────────────────────────────────────┘

EXHIBIT VII *(concluded)*

CERTIFICATION OF APPLICANT

Gulf Teleport, Inc. requests an authorization for construction permit and license in accord with this application, of which all associated exhibits and attachments are a material part.

I certify that the statements in this application are true, complete, and correct to the best of my knowledge and belief and are made in good faith.

Gulf Teleport, Inc.

By _____

Name Olan C. Jones

Title President

Date February 5, 1985

CERTIFICATE OF SERVICE

I hereby certify that I have caused a copy of the foregoing application of Gulf Teleport, Inc. For construction permit and license for transmit-receive earth station in the Domestic Fixed Satellite Service to be served by first-class U.S. mail, postage prepaid, on Hon. Mark White, Governor of Texas, State Capitol Building, Congress Avenue, Austin, Texas 78701, this 11th day of February, 1985.

_____

Emilia L. Govan

■   **Note**   ■

1. For a fuller description of the Gulf Teleport, see C-band application at 4–11.

Authors

Andrew D. Lipman, Esq.
Pepper, Hamilton & Scheetz
Washington, D.C.

Andrew D. Lipman is with Pepper, Hamilton & Scheetz, Washington, D.C., where he specializes in telecommunications and administrative law. Lipman is a graduate of Stanford Law School and University of Rochester, where he was a member of Phi Beta Kappa. He was in the legal honors program at the Department of Transportation and served as a trial attorney in the office of the Secretary of Transportation. He is a member of the Federal Communications Bar Association and the Interstate Commerce Commission Practitioners Association. He is an author of numerous articles on telecommunications, and lectures frequently on the subject.

Emilia L. Govan, Esq.
Pepper, Hamilton & Scheetz
Washington, D.C.

Emilia L. Govan is an attorney with the national law firm of Pepper, Hamilton & Scheetz, Washington, D.C., where she specializes in telecommunications, administrative, and business law. She is involved in the regulatory, business, and financial aspects of teleports, satellite communications, and various radio common carrier services. Govan is a cum laude graduate of the Georgetown University Law Center. She is a member of the Federal Communications Bar Association and co-author of several articles on telecommunications.

Chapter **13**

Using Teleports in Connection with International Communications Services

---

David E. Gourley

COMSAT International Communications, Inc.

Outline

Today's sophisticated antenna farms, if they are to succeed, inherently incorporate a global perspective. The early antenna farms gave birth to the teleport concept, which introduced a new approach to the communications challenges of the 21st century. These challenges are being faced in the United States, along with similar challenges being met in the international arena. This is largely a result of comparable stages of developing communications that have passed through in America, Europe, and Asia.

The questions facing our telecommunications users and service providers overseas are the same ones we are facing in the United States. How well any nation succeeds in coping with its domestic and world issues will inevitably be related to how much attention it gives to its communications requirements. The priorities set for one will play an intrinsic role in the development and, to some extent, the success of the other.

As nations and multinational organizations improve their communications capabilities, international teleports will play an essential role in this process. The process will be driven by an international economy in which international communications gateways will gather and dissiminate information worldwide. Today's teleports provide C-band and Ku-band integrated operations and expanded international coverage, which reveal that the many other existing services are outmoded.

In the simplest of terms, international teleports will connect signals via an international earth station to the INTELSAT global system and on to a final destination. Similar to national teleport functions, the international earth station of a teleport will serve as a hub to transmit and interconnect international communications through multinational networks.

WHY INTERNATIONAL TELEPORTS?

Today, many industrial nations have come to be labeled information-intensive societies. Given recent developments in the telecommunications industry in these countries, it is not surprising to anyone that nations seeking to maintain a strong presence or to acquire a foothold in this industry have challenging tasks. If one considers the development of international telecommunications in the United States, one must consider it in the same context as the development of its international economy.

What's New about International Teleports?

During the past two decades, direct investments by U.S. companies in foreign operations have increased from $49 billion to more than $500 billion, while investments in U.S. operations by foreign-based multinational companies have increased from $7 billion to more than $80 billion. At the same time, the total trade volume, the sum of U.S. Exports and Imports, rose from $34 billion to more than $460 billion in 1980. In the future it should pass the volume of the U.S. corporations operating in foreign countries and the more than 3,000 foreign companies operating in the United States.

The telecommunications industry figures represented about 1.5 percent of the U.S. gross national product (GNP) in 1960, and by 1984 the figures grew to nearly 4 percent. Although data in this area is difficult to obtain, communications make up a significant part of our GNP, and as a result of this industry's growth, the U.S. economy has experienced considerable expansion.

As technology improves, communications users place more stringent demands on communications service and facility providers than they had in the past. They want sophisticated, packaged services to facilitate their international communications needs.

Urban Gateways versus International Gateways

Communications providers are eagerly responding to the customer demands. Today there are a total of 12 operating teleports in the United States, including the teleport on Staten Island in New York, which has an operating international earth station. San Francisco's Bay Area Teleport is under construction, and at least eight others are planned. These teleports include the following: Hawaii Loa in Honolulu; Houston International Teleport in Houston; National Teleport in Washington, D.C.; Ohio Teleport in Columbus; South Star in Davies, Florida; Texas Teleport in San Antonio; and Turner Teleport in Atlanta. These earth stations will supplement the older and larger international gateways. They will not, however, replace them or be a substitute for them.

One reason behind this recent expansion is a large number of newcomers, who for many years paid little attention to communications services and are now beginning to take note. Management is aware of the rapid growth in the international business markets and is taking advantage of new opportunities by combining telecommunications and computer technologies to access and transmit information.

Influences Governing International Telecommunications

By observing business development and the development of communications technology, one can see that each tends to push the other forward. The growth in international commerce and the demand for international communications facilities follow a similar pattern.

Growing numbers of large businesses in the United States and abroad are placing branch or secondary offices in other cities of their homeland, and overseas. The increase and the intensity of international information flow, vis-à-vis voice, data, video, and videoconferencing have also compelled commercial and industrial managers to seek new methods to merge and improve their international communications operations. Each often wants a customized package that will offer flexibility in service applications. They want a system or network that is affordable and can easily be modified to meet growing and future requirements. These demands played a significant role in the initial concept of international teleport operations and consequently in the current international earth station interconnection facilities.

WHO WILL USE INTERNATIONAL TELEPORT COMMUNICATIONS SERVICES?

Communications users with complex services can be better served with facilities as close to the customer's premises as possible, which in turn reduces system performance difficulties that occur over lengthy terrestrial circuits. Although each organization may have requirements unique to its operations, a number of local or regional communications concerns are germane to most communications users operating in the same vicinity. These concerns—technical, policy, development, socioeconomic, and others—collectively generate the need for a new kind of communications facility, which encourages the nation to develop and operate these facilities close to the end user.

Thus the idea of the teleport having a "hubbing" arrangement or earth station farms within close proximity to a commercial metropolis was mothered by technological advancements in telecommunications and fathered by international commercial expansion. In countries where this is in use today, one will find that the need for an international earth station at teleport gateways is a natural expansion of domestic communications needs.

Industries that have begun to utilize international teleport services on a frequent or regular basis include banking institutions, insurance companies, and airline and travel businesses. Broadcasters and publishers are also getting in line for teleport services. These businesses require up-to-the-minute information. For example, a banker might need a quick retrieval on the rate of exchange between two international currencies. A rapid response could give that business manager a substantial advantage over a competitor who has limited access to international communications networks for information retrieval.

The development of international communications technology has enhanced opportunities for financial institutions to pursue multibillion-dollar, multinational projects. New information communications services have made this possible. Teleport services networks are providing affordable and flexible new services through their international communications networks.

NEW COMPETITIVE ENVIRONMENT

Earlier in this chapter, technological advantages that international teleport earth stations brought to the communications users were discussed. In considering another reason behind the development and growth of international earth stations, one finds that the communications industry on the whole has become very competitive. In view of a changing marketplace and a more competitive communications environment, the force driving the international teleports is a rudimentary one, namely, economics. Every day, multinational companies are beginning to see efficiencies in this new approach to their business endeavors.

Organizations having access to state-of-the-art communications systems will naturally have the edge on their competitors who are not as well equipped. When it becomes clear that these new services can streamline operations and remain within a reasonable budget, managers will support the idea of incorporating these services into their operations.

In order to appreciate the significance of today's communications environment, previous communications arrangements should be mentioned. Historically, the telecommunications satellite industry, propelled by a U.S. initiative to create a "global village," put in place an international telecommunications network. The Communications Satellite Corporation (COMSAT), was established to develop a system and achieve this goal. The efforts of COMSAT and the International Telecommunications Satellite Organization (INTELSAT) established a global satellite communications network.

In the 1960s major international gateway earth stations were built in rural areas of the United States to interconnect terrestrial lines with the INTELSAT system. These very large gateways served their purpose well, and they continue to provide quality communications for global traffic.

As demands for more and improved communications services increased, technology provided an alternative in services. A decision was made to modify the international satellites, thereby making it possible to build smaller earth stations to accommodate as much, and even more, traffic than their predecessors. Some of the smaller earth stations became the prototype for today's international earth stations.

Prior to the introduction of teleports, international business customers (most of whom were located in major metropolitan business centers) had their international communications services routed through remote international gateway earth stations. The commercial international communications earth station facilities were and still are located in rural areas throughout the United States. They provide international services via INTELSAT. Today both rural and urban gateways use the INTELSAT system.

New digital services are coordinated through INTELSAT with the intention of utilizing both international teleports and major gateway earth stations. They became a reality when the INTELSAT Board approved a wide range of new communications services, which INTELSAT refers to as International Business Services (IBS). INTELSAT, which represents 110 country members, has thereby created a positive global welcome to communications representatives who seek to achieve technological harmony in the international arena. INTELSAT's initiative has also served as a key to paving the way for communications facilitators to work through national Postal Telephone and Tele-graph (PTT) administrations to acquire necessary licenses and approvals. COMSAT, as U.S. signatory to INTELSAT, has provided and will continue to provide these services for its U.S. customers and carriers. The international business services are also known by COMSAT International's digital network service, Digital Express.

Countries that have begun efforts to participate in the new communications services in addition to the United Kingdom include Canada, The Netherlands, West Germany, Italy, France, Sweden, Denmark, Norway, Mexico, Switzer-land, Belgium, Japan, Singapore, Australia, and Hong Kong.

COSTS

International communications services and facilities bear considerable cost. The customer is always looking at affordability, and some industries more than others require absolute assurance of (1) no system failures, and (2) a secure transmission signal. To achieve these goals communications providers have placed substantial resources in developing technology in both these areas.

In the field of communications, costs often decrease as technology improves. The teleport concept is no exception. It responds to a cost-conscious environ-ment by bringing a variety of new, cost-efficient, value-added telecommunica-tions services closer to the customer. Substantial long-distance terrestrial costs have been reduced if not eliminated. Today's teleports, and those on the drawing board, are all located within urban business centers or just outside of them and are thus nicknamed "urban gateways." It is reasonable to believe that most major international municipalities in the United States and abroad that boast of aggressive business markets will have their own teleports by the mid to late 1990s.

Another reason for the acceptance of U.S. teleports could be related to recent deregulation trends. As leaders in bustling industrial centers cope with the changes brought about by deregulation in communications, they are also taking a close look at the economics of having an urban gateway—that is, the idea of merging new technologies of telecommunications and computers as a means of lessening the cost of voice, data, and other communi-cations services.

For the long-distance users, particularly the international users, the accessi-bility of an antenna within a short distance of the office generally proves to be more economical than traveling terrestrially for long distances, especially if one has a high volume of constant traffic. Another advantage teleports offer customers is lower rates for high bit rate service on the satellite circuits. These advantages are points supporting urban gateways as economical alterna-tives that offer flexibility of services to the end user.

It is estimated that teleports could reduce end user's communications costs by 20 to 40 percent. It is difficult to estimate costs, however, since technology, location, volume of traffic, and local regulations vary. International requirements and regulations determined by governing telecommunications administrations must also be taken into consideration. This latter constraint is a major wild card that makes premature predictions impractical. One can look at some alternatives being considered that will ultimately influence the timetable for development of an international teleport network.

COORDINATION WITH FOREIGN ENTITIES

Although the required technology to either send or receive signals long distance has become rather straightforward, the timetable is subject to the emphasis consequently being placed on political and regulatory questions. In other words, the rapidity with which urban gateways grow, particularly on an international level, is dependent on how quickly the political constraints can be removed and the problems resolved.

For example, Dow Jones & Company, Inc., a leading U.S. newspaper publisher, is already enjoying numerous benefits from printing its paper remotely via satellite across the continental United States. Consequently, its domestic network has, for the first time, truly produced a national newspaper.

The main benefit that this new approach has offered to Dow Jones is being appreciated in substantial cost savings to the company. In addition, the publisher is remotely printing a limited number of papers overseas and has indicated an interest in escalating this endeavor once the more bureaucratic and regulatory constraints have been resolved. Coincidentally, newspapers in France, Greece, Italy, and Germany have expressed an interest in remotely printing their major daily publications abroad.

In looking at the issue of international coordination, one should mention the safeguards taken to ensure that the same high standards demanded in other communications activities continue throughout the development of urban gateways. The global superintendent, whose job it is to ensure the coordination of the grand scheme, is INTELSAT. The responsibility lies here because of the INTELSAT Agreement (a treaty among the countries that participate in INTELSAT) and the Operating Agreement (an agreement among the entities—signatories—designated by the governments of those countries to participate in INTELSAT operations). Both agreements have been in effect since 1973.

Procedurally, arrangements for coordination of teleport services with foreign ends are handled through the INTELSAT signatories. Communications Satellite Corporation (COMSAT), as U.S. signatory, fulfills this role for U.S. customers. Connectivity, access to the space segment and end-to-end costs, and so forth are all arranged by the signatories with the cooperation of INTELSAT, which designates frequency allocations for international earth stations. On the other hand, teleports are expected to provide regular communications services and a number of new communications services using other satellite systems, for example, Eutelsat and Telecom in Europe.

Other countries sharing a mutual boarder may follow the example of Canada and the U.S. Currently, each country is bilaterally using its domestic satellites rather than INTELSAT to carry traffic between both countries. Consequently, international traffic will follow regulatory procedures of the countries' domestic rather than international systems. Once again, it is apparent that a major concern is one of administration, not technology.

■   SUMMARY   ■

If one were to observe the path teleports seem to be taking from a marketing perspective, one could quickly conclude that teleports are growing rapidly and that they are keeping pace with the growth of other communications services. Marketplace demand created them; they were not a clever, blind invention. They are a natural outgrowth of world economics in an information-intensive environment that is being led by international decision makers who demand excellence.

As technology continues to develop, competition continues to climb, customer demands increase, and service delivery maintains a balance, this network is bound to be the right combination, which will serve as the foundation for an international teleport network. Many countries recognize its virtues and have begun to tackle some of the administrative constraints that must be resolved before they can begin to implement their plans. Although each country's administrative communications operations differ, this author looks ahead to significant growth in the development of an international teleport network.

Author

David E. Gourley
Vice President—Marketing and Sales, COMSAT International
Communications, Inc.

David E. Gourley is vice president of marketing and sales for COMSAT International Communications, Inc. In this position, Gourley is responsible for the establishment of new business plans, new communications services, and market development for COMSAT International.

He joined COMSAT in 1976 and most recently served as vice president of business development for COMSAT World Systems Division. Prior to this he had been director of sales and market development for World Systems since 1981.

Gourley's communications experience dates back to 1952 when he worked for the Air Traffic Division of the Civil Aeronautics Agency. There he assisted in the design and implementation of a computer-assisted air traffic control system. In 1962 he joined UNIVAC, serving as technical manager, and from 1968 to 1976 he held several positions with Data Transmission Company.

A native of Pennsylvania, Gourley and his family currently reside in Fairfax, Virginia.

Chapter 14

Connecting the High Tech Building through Satellite Networks

John N. Lemasters
Continental Telecom Inc. (Contel)

Albert F. Caprioglio
Communications Systems Development

William J. Rahe, Jr.
American Satellite Company

George R. Welti
Consulting Engineer

Outline

Source: Reprinted from Alan D. Sugarman, Andrew D. Lipman, and Robert F. Cushman eds., *High Tech Real Estate* (Homewood, Ill.: Dow Jones-Irwin, 1985), © Dow Jones-Irwin, 1985.

Previous chapters described teleports and their impact on the intelligent city. This chapter discusses using satellite technology for the interconnection of high tech buildings and teleports in these cities to meet their information transfer requirements. These requirements range from low-speed data through voice and video conferencing up to high-speed data transfer.

In order to adequately address the subject, this chapter analyzes the various aspects of establishing a satellite telecommunication system among high tech buildings, including a basic description of the satellite transmission system, specific examples of existing satellite networks, the actual position of the satellite, and how this positioning along with other geographical or environmental issues affect the specific design of the high tech building.

WHY SATELLITE COMMUNICATIONS?

In discussing the use of satellites to transfer information between high tech buildings, one of the questions that arises is "Why satellite?" Why not terrestrial systems, such as fiber optics or digital microwave? This type of question implies an either-or situation. In reality, most networks will utilize a combination of technologies in the design of the system, balancing the advantages of each against the specific requirements of the situation.

The main advantage of satellites is the ability to transmit signals anywhere from the continental United States to the satellite and have the satellite re-transmit these signals from space to earth. The signals can then be received at any point on earth that lies within the region covered by the satellite's "down beam" or "footprint." These footprints can include large regions, such as Canada, the Continental United States, Mexico, Colombia, and Brazil. In some cases offshore points, such as Hawaii and Puerto Rico, are also included. Figure 14–1 depicts the Ku band footprint of ASC I.

The coverage provided by the satellite beams enables satellites to provide unique wide-area broadcasting capabilities for such applications as television distribution to over-the-air and cable broadcasters, direct-to-home television broadcasting, and private network video teleconferencing. This wide-area point-to-multipoint capability is unique to satellites and cannot be matched by terrestrial radio (except for voice-only broadcasting) or by wireline telephone carriers.

In addition to providing TV and teleconferencing, the satellite broadcast mode is also used for such business applications as facsimile transmission and information dissemination. Such newspapers as *The Wall Street Journal, The New York Times,* and *USA Today* are printed simultaneously in many cities throughout the country from reproducible master copies broadcast to the printing plants via satellite. Stock market quotations and news agency

FIGURE 14–1
ASC–1 EIRP Contours

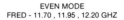

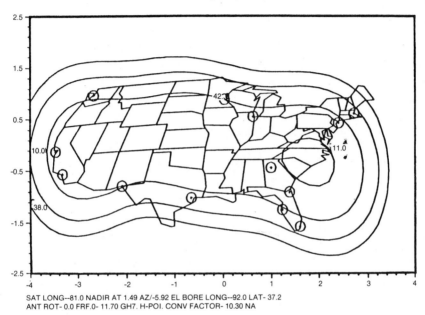

SAT LONG--81.0 NADIR AT 1.49 AZ/-5.92 EL BORE LONG--92.0 LAT- 37.2
ANT ROT- 0.0 FRF.0- 11.70 GH7. H-POI. CONV FACTOR- 10.30 NA

copy are disseminated to broadcasters, cable operators, and business sub-scribers via satellite. In some cases interactive services are offered in which the customers can request specific information from a central data base. Requested information can be received via satellite with a small-dish receiver (0.6 meters) and can be displayed on a monitor or printed.

Government and business users of telecommunications are served by satellite networks that include on-premise transmit/receive earth stations. Another arrangement for private business networks makes use of carrier-owned earth station and central office facilities usually located in major metropolitan loca-tions—New York City, Chicago, Los Angeles. Access from customer locations to these facilities may be provided either through facilities leased from the local telephone company or via alternative means, such as microwave radio links or fiber optic cables, that bypass the local telephone facilities.

Point-to-point and multipoint networks are employed by users for private voice and data communications and for teleconferencing. The advantages of satellite over interstate terrestrial networks include low cost, high availability, and high quality.

The main reason for these advantages is the differing geometrical characteristics of satellite and terrestrial communications networks (e.g., connectivity, signal path length, and ground distance between end points of a path), which lead to fundamental differences in their technical and cost characteristics.

Examples of connectivity maps for terrestrial and satellite networks are shown in Figure 14–2. Major land routes (such as those of the AT&T long-distance network) are depicted in Figure 14–2a; the full network includes hundreds of additional routes and about 160 switching centers. A typical end-to-end signal path traverses between 4 and 10 switching centers. A private business network can comprise a large number of such paths.

By contrast, the connectivity map for a corresponding satellite network is very simple. As seen in Figure 14–2b, all satellite earth stations "see" the satellite, and vice versa. All signals sent up to the satellite are rebroadcast to all earth stations. A business network can be established simply by installing on-premise earth stations at all locations. Adding one new location to an existing network requires the addition of only one new earth station.

When compared with terrestrial signal path length, satellite path lengths are essentially distance insensitive, although much longer, since the signal is routed through a satellite located approximately 22,300 miles above the equator.

The contrast between distance-sensitive terrestrial path lengths and distance-insensitive satellite path lengths is reflected in the costs of terrestrial and satellite circuits. This is illustrated in Figure 14–3 which compares private-line voice circuit tariffs quoted by terrestrial and satellite carriers in June 1984. The figure shows that tariffs for terrestrial circuits are indeed quite distance sensitive, and those of satellite circuits are much less so. For this reason, local networks rarely employ satellites; regional or national networks represent typical applications for satellite services.

In applying the general concept of telecommunications networks against a specific situation, the issues of types of applications supported, traffic volumes, connectivity requirements, quality, and costs all have to be considered. The network that results from the subsequent design is a single-point optimization and must be modified based on changing conditions.

SATELLITE COMMUNICATIONS NETWORKS

Over the past few years, the telecommunications environment has undergone some radical changes, the effects of which have significantly altered companies' approaches to networking. Of the various alternatives, three types of satellite networks are emerging. They are dedicated private networks, shared private networks, and virtual private networks. An example of each of these types of networks is given below. Communication network services

FIGURE 14–2

A. Terrestrial network (simplified)

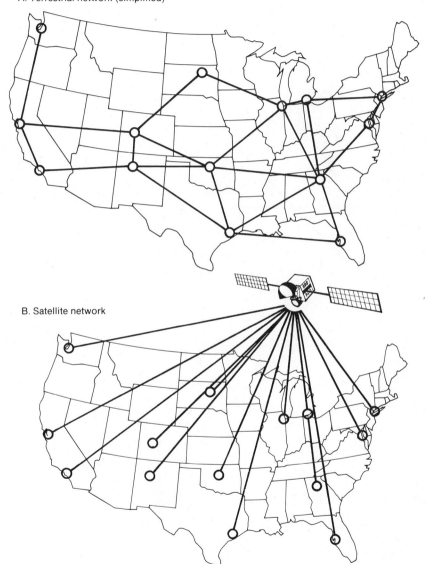

B. Satellite network

FIGURE 14–3
Tariffs for Private-Line Voice Circuits

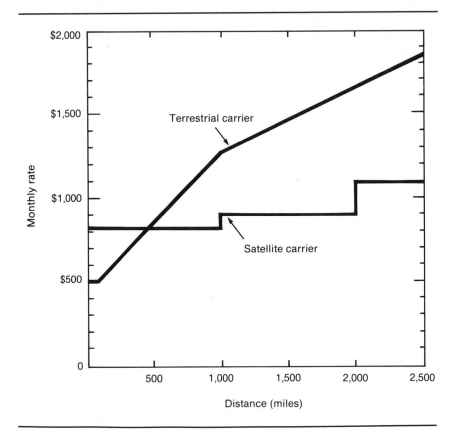

for the business users cited are provided by American Satellite Company and other companies.

Dedicated Private Networks

A dedicated private network is the most common type of network in today's environment and is the most likely network to be accessed or utilized from high tech buildings. These networks generally have between three and eight major nodes or centers and are distinguished by having both the switching function and the transmission paths dedicated for the customer's sole use. An example of this type of configuration is the General Dynamics Integrated Digital Network. The General Dynamics Network is configured to meet corporate requirements for voice and data services and consists of six customer premise nodes: Pomona, California; San Diego; Sterling Heights, Michigan; Fort Worth; Creve Coeur, Missouri; and Groton, Connecticut. As shown in Figure 14–4, the system utilizes the DEX 400S as its primary switching vehicle. The network consists of more than 250 intermachine trunks (IMTs); 83 full-duplex data circuits (ranging from 2.4 kbps through 230.4 kbps); 719 access circuits; and 278 WATS, FX, and DDD circuits.

The major benefit of this type of network is the flexibility of having dedicated resources colocated at the customer premise. This type of arrangement allows the customer to quickly reconfigure or to rebalance the network as changes in requirements dictate. A major drawback is the requirement for significant traffic volume to operate cost effectively.

FIGURE 14–4
General Dynamics Dedicated Network

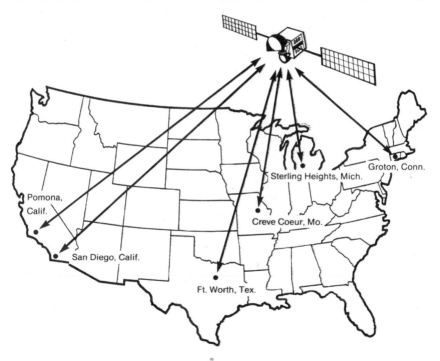

● General Dynamics Locations
All nodes utlize DEX 400S Switches
Connectivity between all locations is 512 kbps or multiples thereof.

Shared Private Networks

A second type of network that is gaining popularity is the shared private network. The main distinction between this network and the dedicated private network is that some of the switching and/or transmission resource is not located on the customer's premise but at a centralized location, which aggregates different network requirements to achieve the economies of scale that result in a cost-effective system.

An example of this type of network is the Federal Express network. The Federal Express network consists of four nodes: one shared (San Francisco) and three dedicated (Colorado Springs, Memphis, and Somerset, New Jersey)

FIGURE 14–5
Federal Express Shared Network

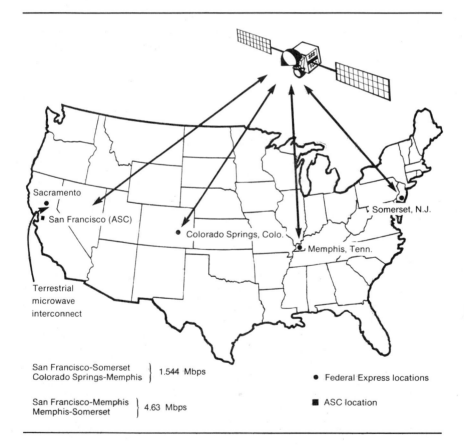

Sacramento

San Francisco (ASC)

Somerset, N.J.

Colorado Springs, Colo.

Memphis, Tenn.

Terrestrial
microwave
interconnect

San Francisco-Somerset
Colorado Springs-Memphis   } 1.544 Mbps

San Francisco-Memphis
Memphis-Somerset   } 4.63 Mbps

● Federal Express locations

■ ASC location

as illustrated in Figure 14–5. The system utilizes Collin's digital switches for the voice traffic and utilizes more than 300 IMTs and the data capability of nine 56 kbps point-to-point, full-duplex circuits.

This type of network provides many of the same advantages as does the dedicated private network. Where the traffic requirements and other considerations warrant, dedicated customer premise equipment is provided. However, where the traffic requirements are insufficient to warrant dedicated locations, additional flexibility is provided through the utilization of shared locations. The trade-off is some loss of direct control over that portion of the system.

Virtual Private Networks

One of the emerging network types gaining ascendancy is the virtual private network. The main feature distinguishing this type of network from the others is that all of the locations at which a customer terminates traffic are shared. They not only are shared in the sense that multiple customers are terminated at the centralized location, but also with respect to the switching and/or transmission paths utilized. Although components are shared, they appear not to be for the user, and for that reason, the term *virtual* is used. The only items that are truly dedicated to a particular customer are the access circuits that connect the customer's location to the centralized location. This sharing significantly improves network flexibility while minimizing the capital risk to the network user.

In order to maintain the required privacy and to keep the various networks separate, the switches and transmission facilities are "partitioned" by modifying the switch software. An example of such a system is the planned Sears network.

The concept behind this network is to connect the various Sears locations to approximately nine centralized sites: New York; Washington, D.C.; Miami; Atlanta; Chicago; Dallas; Los Angeles; San Francisco; and Seattle. At these locations, Sears will use a partition on a Digital Switch Corporation's DEX 400/400S and will have the option of either dedicated or shared Intermachine Trunks (IMTs).

This type of network is significantly different from either the dedicated private network or the shared private network, since only the access facilities are dedicated to a specific user. Implicit is a loss of direct control by the users over their network. In order for this service to be accepted by the users, the network manager must provide detailed information on a timely basis to the users and must also be extremely responsive to user change and trouble reporting requirements.

Loss of control and greater dependence on the network manager due to the sharing of lines and equipment are weaknesses of this type of network, but the strengths arise from those same traits. Because of the extensive amount

FIGURE 14–6
Sears Virtual Private Network

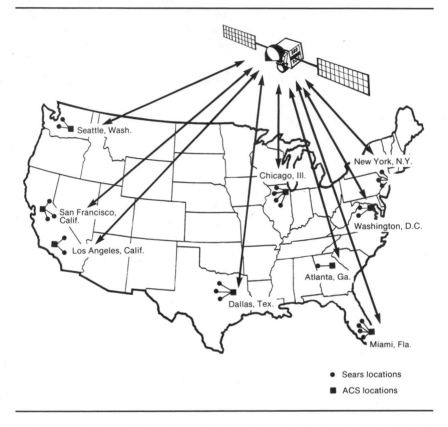

● Sears locations
■ ACS locations

DEX 400S Switches at all ASC locations; software partition, utilized. System operates at 60 to 120 Mbps.

of resource sharing, virtual private network services can accommodate the necessary traffic volumes to cost effectively provide for a wide range of services, such as protocol conversion, switched data services, and packet switching. Additionally, this type of network does not require the user to provide dedicated direct trunks between the switching vehicles. Since the connectivity factor[1] is no longer an issue, the users are not limited to only a few centralized locations for their network nodes but can expand their networks to 20 or more centralized locations or nodes. This allows companies to cost effectively connect a much larger number of their locations with advanced communications services.

INCORPORATING A SATELLITE NETWORK INTO THE HIGH TECH BUILDING

After it has been determined that a satellite-based network is cost effective and meets the general requirements of connectivity, types of applications, traffic volume, and quality, a specific network design must be generated. As part of that design effort, a determination must be made of the best way to integrate the high tech building into the network. This integration can consist of:

A rooftop satellite access antenna with a diameter between 3 and 7 meters, depending on the satellite network the user intends to access.

A terrestrial microwave antenna that interconnects a remotely located earth station (especially if C band is used) with the user's building.

Local area networks interconnecting high tech buildings within an industrial complex (i.e., fiber optic cable, coaxial cable).

A fiber optic cable connecting to a teleport or off-site antenna.

In determining which of these is the optimum integration technique, the following areas must be investigated: regulatory, orbital arc look angles, microwave line-of-sight and interference, and building asthetics.

Regulatory Impact

FCC approval. Provision for satellite network access involves more than acquiring the requisite technical components. Various regulatory requirements must be met, such as (1) obtaining a license from the FCC, (2) obtaining local community approvals (depending on zoning regulations), and (3) achieving frequency coordination that is successful in that it demonstrates compatibility with both existing systems and those terrestrial and satellite communication systems whose applications have been filed with the FCC.

A license must be obtained from the FCC to build and operate a transmit/receive earth station or terrestrial microwave stations. Constraints on earth station and microwave relay configuration and performance are imposed by the FCC; station technical characteristics must meet FCC standards and various environmentally imposed microwave power flux density radiation constraints.

An important part of the process of obtaining FCC approval is demonstrating compatibility with existing or planned (approved) transmission/reception systems. A frequency search of existing or planned communication systems, conducted by one of the firms specializing in this process, is the first step in this demonstration of frequency compatibility/coordination. Location of the satellite antenna is critical to avoid interference with existing systems; acceptable locations within the complex should be identified early in the development process. This is especially critical for C band satellite access, since current C band terrestrial microwave interconnections utilize the same frequency band as do communications satellites. Interference standards or preliminary frequency search results may show that an originally planned location is unacceptable; if so, alternative locations within the complex should be found. If this is difficult or impossible, shield walls may be used to reduce interference to acceptable levels, but only if not in a direct line with existing terrestrial communications systems.

Local zoning. If construction of the high tech building or complex involves zoning changes within the local community, requirements for zoning approvals should include the total configuration, especially if an earth station is incorporated into the high tech building design or if a multiple antenna earth station (teleport) is planned to serve the industrial complex. Height limitations may be imposed, restricting antenna placement, as well as constraints on the number and size of antennas. Aesthetic appeal is a major consideration within established communities, coupled with strong concern over radiation hazards. Many local communities require demonstration of minimum environmental impact—for example, that the radiation from a satellite transmissions antenna or from a microwave relay antenna presents no danger to the local community. The problem is compounded by the absence of universally accepted standards.

Most communities don't have specific zoning ordinances or environmental guidelines that adequately cover satellite earth stations. If an antenna is part of a building function, it may be included within certain zoning regulations. Usually not included are maximum radiation levels and constraints on location and pointing of earth station antennas. Some communities have passed ordinances that govern the number, location, and placement of satellite antennas, including maximum levels for power flux density. Also imposed is periodic verification of not-to-exceed limits. High tech building developers should work closely with community representatives and communications experts in the initial stages of program/complex development to ensure the multiple satisfaction of goals and avoid later redesign or modification when approval is sought.

Orbital Arc Look-Angles

As discussed above, the integration of network access capability into a high tech building or complex must take into account a number of factors, which include frequency coordination, terrestrial interference, and zoning requirements. The location of satellite access or microwave interconnect antennas is affected by orbital arc look-angle requirements, existing interference from terrestrial microwave communications links, and microwave interconnect line-of-sight (if access to a remotely located satellite station is required).

The antenna look-angle—antenna-pointing azimuth and elevation angles to the satellite location—is a function of both satellite location within the U.S. domestic orbital arc and location of the high tech building in the United States. An example of these satellite angles as they relate to earth station location is illustrated in Figure 14–7. Representative azimuth and elevation angles from selected locations within the United States to satellite locations with the U.S. domestic orbital arc are also shown.

A minimum antenna elevation angle of 10 degrees is necessary to achieve proper communications performance; higher elevation angles are desired to minimize blockage, atmospheric and rainfall attenuation, and interference

---

FIGURE 14–7
Earth Station Azimuth and Elevation Look-Angles

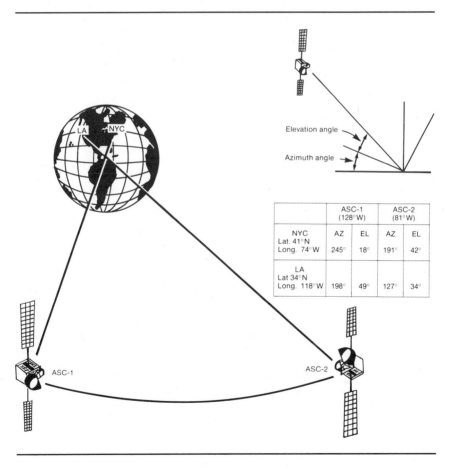

	ASC-1 (128°W)		ASC-2 (81°W)	
	AZ	EL	AZ	EL
NYC Lat. 41°N Long. 74°W	245°	18°	191°	42°
LA Lat 34°N Long. 118°W	198°	49°	127°	34°

from terrestrial communication services. Lowest elevation angles are experienced from earth station locations in the northernmost tips of the United States (i.e., Maine and Washington) and to satellite locations at the eastern and western extremes of the U.S. domestic orbital arc (approximately 62° west to 146° west longitude—refer to Table 14–1).

Microwave Line-of-Sight and Interference

A high tech building developer who is considering the use of satellite networks to fulfill business communications requirements within a complex must make a choice on the access method to be employed for establishing an interface with the satellite network. As mentioned previously, the earth station may be located either on or off the user's premises. The determining factor in this choice is usually cost. If a dedicated (on-premise) earth station is employed, its cost is borne by the user—either directly or indirectly. By sharing the use of a centrally located earth station with others, this cost can often be reduced substantially. On the other hand, use of such a remote earth station entails either access-line charges imposed by the local telephone company or the cost of a microwave interconnect.

The siting of a microwave antenna to interconnect a high tech building with a satellite earth station involves conflicting requirements. On the one hand, location at a point of maximum visibility to the remotely located earth station is desired, as well as maximum protection from physical construction/obstruction. On the other hand, that same location is most vulnerable to interference from existing terrestrial microwave communication links. An illustration of microwave interconnect line-of-sight requirements is given in Figure 14–8. If at all possible, the use of existing terrain and buildings to increase blockage of unwanted signals is desirable. Incorporation of the antenna into the building structure, using architectural features to block terrestrial interference, is also appropriate in many cases. Aesthetic considerations must be taken into account, not only from a business environment standpoint, but also in response to community requirements. Local zoning ordinances may impose constraints on the appearance and placement of network access antennas. The need for municipal design approval may levy additional requirements on system configuration.

Building aesthetics. Overall high tech building appearance may be improved by masking access antennas with some type of existing terrain shielding—for example, landscaping and trees may be used as long as they are not placed within the antenna look-angle arc. A rough rule is to maintain an intrusion-free clearance cylinder equal to the antenna diameter and concentric with the antenna bore axis. A potentially more rewarding approach to aesthetic improvement is to incorporate access antennas into the building

TABLE 14-1
Existing (or planned) Orbital Arc Assignments—62° West to 146° West Longitude

Longitude	Satellite Name	Longitude	Satellite Name
62.0	SATCOM 7 (C)*§	101.0	SBS 4 (Ku)‖
62.0	SBS-6 (Ku)†§	103.0	GSTAR 1 (Ku)
64.0	ASC-3 (H)††§	104.5	ANIK D (C)
67.0	SATCOM 6 (C)§	105.0	ANIK C-2 (Ku)(USCI)
67.0	RCA C (Ku)§	105.0	GSTAR 2 (Ku)§
69.0	SPACENET 2 (H)	109.0	ANIK B (H)
71.0	GALAXY A (Ku)§	111.5	ANIK D-2 (C)§
72.0	SATCOM 2R (C)	112.5	ANIK C-2 (Ku)§
73.0	WESTAR A (Ku)§	113.5	MORELOS A (H)§
74.0	GALAXY II (C)	114.0	ANIK A-3 (C)
75.0	COMSAT GEN A (Ku)§	116.5	MORELOS B (H)§
76.0	COMSTAR II/III (C)	117.5	ANIK C-3 (Ku)
77.0	FEDEX A (Ku)§	120.0	SPACENET I (H)
79.0	WESTAR III (C)§	122.0	SBS-5 (Ku)§
79.0	WESTAR II (C)‖	122.0	Unassigned (C)
79.0	MARTIN MAR-A (Ku)§	122.5	WESTAR 5 (C)‖
81.0	RCA B (Ku)§	124.0	WESTAR 5 (C)§
81.0	SATCOM 4 (C)§	124.0	FEDEX B (Ku)§
83.0	SATCOM 4 (C)‖	125.0	TELSTAR 303 (C)‖
83.0	ASC-2 (H)§	126.0	MARTIN MAR-B (Ku)§
85.0	RCA A (Ku)§	126.0	TELSTAR 303 (C)§
85.0	TELSTAR 302 (C)§	128.0	ASC 1 (H)
86.0	TELSTAR 302 (C)‖	130.0	GALAXY B (Ku)§
87.0	SPACENET 3 (H)§	130.0	SATCOM 3R (C)§
89.0	Unassigned (H)	131.0	SATCOM 3R (C)‖
91.0	WESTAR 3 (C)‖	132.0	WESTAR B (Ku)§
91.0	WESTAR 6S (C)§	132.0	GALAXY I (C)§
91.0	SBS-4 (Ku)§	134.0	GALAXY 1 (C)‖
93.0	FORD II (H)§	134.0	Unassigned (C)
93.5	GALAXY 3 (C)‖	134.0	COMSAT GEN-B (Ku)§
95.0	GALAXY 3 (C)§	136.0	SPACENET/GSTAR (H)§
95.0	SBS 3 (Ku)	138.0	SATCOM 1R (C)§
96.0	TELSTAR 301 (C)‖	139.0	SATCOM 1R (C)‖
97.0	SBS 2 (Ku)	140.0	GALAXY 4 (C)§
97.0	TELSTAR 301§	142.0	AURORA (C)§
99.0	SBS 1 (Ku)	143.0	SATCOM 5 (C)‖
99.0	WESTAR 4 (C)	144.0	WESTAR 7 (C)§
101.0	FORD I (H)§	146.0	AURORA (C)§

Note: The locations are based on the FCC order of July 25, 1985.
* C band.
† Ku band.
‡‡ Both C band and Ku band.
§ Not yet at this location.
‖ Moving from this location.

FIGURE 14–8
Microwave Interconnect Line-of-Sight Illustration

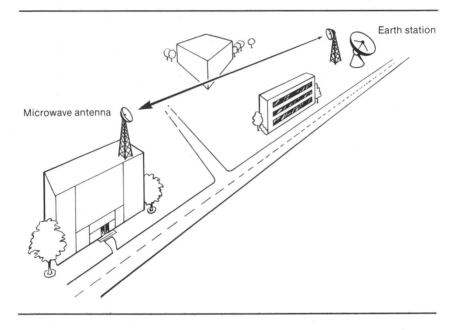

or complex design configuration. Illustrative examples of the approach for direct satellite access from high tech buildings themselves are given in Figure 14–9.

If the building location is such that unobstructed orbital-arc visibility is possible from the ground level, then the building design could be such as to "encompass" a good portion of the antenna, potentially achieving the dual objectives of visual screening from public view and shielding from electromagnetic interference. Alternatively, if ground-level visibility is not possible, it may be necessary to place antennas on the roof. Architectural design features that integrate the antenna visibly with the building could potentially produce more acceptable results than simply "tacking on" an antenna to a building whose design concept had not made provision for one. Since terrestrial microwave antennas have no need for orbital-arc visibility, they are more easily incorporated into building design.

With a remotely located antenna "teleport" serving an industrial complex, more opportunities exist to produce a configuration aesthetically acceptable to the developer and the community: flexibility in earth station placement and the ability to incorporate natural and artificial screening are examples.

Surrounding terrain in a suburban setting may provide natural screening for both visual and electromagnetic interference purposes. Where conditions warrant, earth berms can be constructed or, conversely, antenna placement may be below ground level, permitting unobtrusive shielding. Figure 14–10 illustrates the concept of a centrally located antenna teleport serving an industrial complex.

FIGURE 14–9
Direct Satellite Access from Building

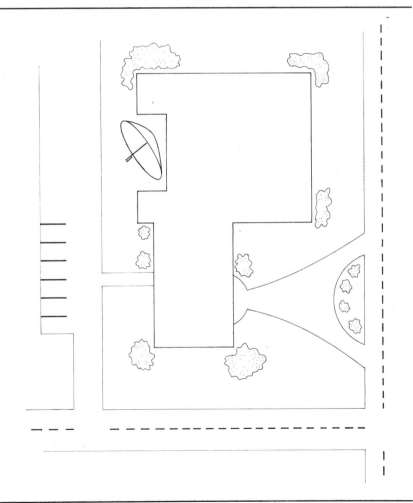

FIGURE 14-9 (*concluded*)

High Tech Building—Evolution of the Satellite Network

Three examples of satellite networks into which high tech buildings could be incorporated were described earlier in this chapter, followed by a discussion of some of the more important considerations in incorporating the building into a network. It is appropriate to close this chapter with a short summary of the next step in the satellite networks evolution: the integrated services digital network, ISDN.

The example networks described have several major areas in common:

FIGURE 14–10

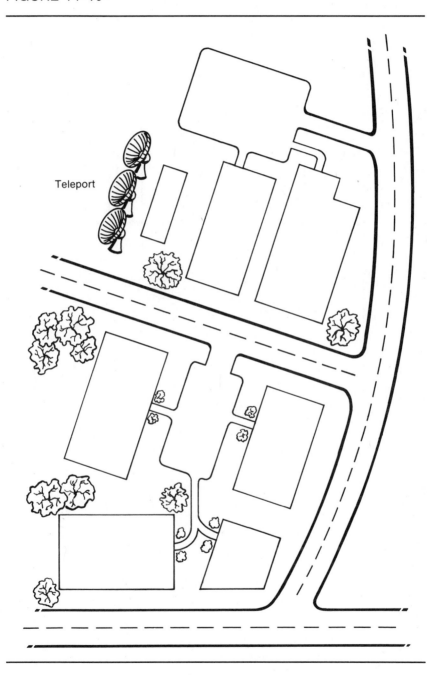

Teleport

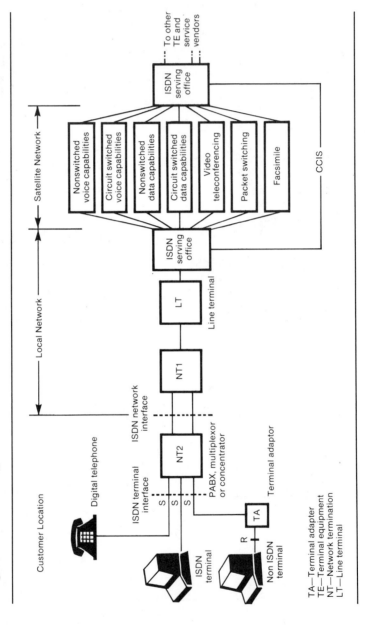

FIGURE 14–11
Integrated Services Digital Network

Customer Location

Digital telephone

ISDN terminal
interface

ISDN network
interface

Terminal adaptor

S

S

S

NT2

NT1

LT

ISDN
serving
office

Local Network

Satellite Network

PABX, multiplexor
or concentrator

Line terminal

ISDN
serving
office

ISDN
terminal

Non ISDN
terminal

R

TA

Nonswitched
voice capabilities

Circuit switched
voice capabilities

Nonswitched
data capabilities

Circuit switched
data capabilities

Video
teleconferencing

Packet switching

Facsimile

CCIS

To other
TE and
service
vendors

TA—Terminal adapter
TE—Terminal equipment
NT—Network termination
LT—Line terminal

(1) digital transmission paths that carry both voice and data traffic, (2) digital switching with stored program control, and (3) the necessity to meet the company growth requirements. The natural outgrowth of these networks is the ISDN.

The ISDN objective is to provide end-to-end digital telecommunications services supported by a common network. Therefore, such services as switched and nonswitched voice, switched and nonswitched data, video teleconferencing, facsimile, and packet switching traffic would all be serviced via the same network.

This does not mean that all of these services are aggregated via one common "black box." Instead, these services will utilize the necessary equipment that translates them into a standard digital format. These formats can then be integrated and transmitted over a common digital transmission path (as illustrated in Figure 14–11).

The basic features of the ISDN are:

Sharing of digital switches and digital transmission paths among different/multiple services (e.g., voice, data, video teleconferencing, facsimile).

Common Channel Interoffice Signaling (CCIS) for transportation of network information and signaling messages.

Intelligence at the switching nodes to store and to process information for a wide range of enhanced information and telecommunications services.

Standardized network interface and protocols.

The ultimate goal of the ISDN is to economically provide a wide range of basic and enhanced services designed to meet the user requirements through the use of shared equipment and standardized interfaces/protocols.

■ CONCLUSION ■

The preceding has described how satellite technology can be used to meet business information transfer requirements by integrating high tech buildings into satellite-based telecommunications networks. The advantages of satellite communications were discussed, as were examples of existing business satellite networks. Constraints and guidelines for incorporating satellite network access into high tech buildings were addressed, together with a summary of what the future will bring in digital telecommunications systems which will provide voice, data, and video services in one integrated network.

■   NOTES   ■

1. In designing a fully connected network, the number of dedicated direct trunk groups is dependent upon the number of nodes. As the number of network nodes (n) increases, the number of trunk groups increases at the rate of $n(n-1)/2$. Therefore, for a network with 3 nodes, the number of trunks is $3(3-1)/2$, or 3; for 10 nodes, the number of trunk groups would be $10(10-1)/2$, or 45, and for 20 nodes, the number of trunk groups required becomes 190!

Authors

John N. Lemasters
President and CEO
Continental Telecom Inc. (Contel)
Atlanta, Georgia

John N. Lemasters became president and CEO of Continental Telecom Inc. (Contel) in March 1985. Before that, he had been president and CEO of American Satellite Company since January 1984. American Satellite, a unit of Contel, has been a leader in supplying long-distance communications to the business community. Prior to joining American Satellite, Lemasters was senior vice president, sector executive at Harris Corporation, where he was responsible for the communications segment of the business. He joined Harris Corporation as an engineer after graduating from Georgia Institute of Technology in 1958 with a B.E.E.

Albert F. Caprioglio
President
Communications Systems Development
Washington, D.C.

Albert F. Caprioglio provides advanced communications systems consulting services to satellite- and terrestrially based telecommunications companies. He was formerly vice president, Advanced Systems, for American Satellite Company, responsible for long-range systems planning and development. Caprioglio has considerable experience in space communications systems, including responsibility for the Galaxy program at Hughes Communications, Inc., where he was vice president. A graduate of New York University (electrical engineering), he received a M.S. degree in electrical engineering from Princeton University.

William J. Rahe, Jr.
Assistant Vice President, Marketing Development
American Satellite Company
Rockville, Maryland

William J. Rahe, Jr., assistant vice president, Marketing Development, for American Satellite Company, is responsible for long-term business planning, new product development, and acquisitions. Rahe has considerable experience in telecommunications, including responsibility for the CNS-B Service Program at Satellite Business Systems, where he was a director prior to joining American Satellite Company. A graduate of the University of Illinois (electrical engineering), Rahe received his MBA from Loyola University in 1975.

George R. Welti
Consulting Engineer
Leesburg, Virginia

George R. Welti is a consulting engineer. Most recently, he served as a senior staff scientist at American Satellite Company, where he was responsible for advanced system design covering the 5- to 15-year future time frame. His scope includes both the space and the terrestrial segments of telecommunications service facilities. Welti has authored many professional papers on telecommunications. He received M.S. degrees in both mechanical and electrical engineering from the Massachusetts Institute of Technology in 1948 and 1955.

Chapter 15

Teleports and Bypass: Opportunity or Threat to Local Telephone Companies?

Joseph S. Kraemer
Touche Ross & Co.

Outline

THE NATURE OF BYPASS

One of the most controversial subjects in telecommunications today is bypass. The Federal Communications Commission (FCC) has used bypass to justify its access charge decision. Bell Operating Companies (BOCs) and some independent telephone companies take bypass as a given and request "a level playing field." State regulators, in turn, tend to believe that bypass is: (1) not a problem, (2) created by telephone companies as an excuse for rate increases, or (3) relevant to states other than the one in which they preside.

In order to determine the extent of bypass, Touche Ross & Co. interviewed approximately 2,000 of the largest end users of telecommunications throughout the United States.[1] Approximately 5 percent of these users were public agencies, such as cities, counties, and school districts. The remaining 95 percent were large business customers. The interviews began in December 1982 and are still in progress.

For purposes of our studies, *bypass occurred when voice or data traffic was originated and/or terminated without using an exchange carrier's local loops.* The technologies employed to bypass a local telephone company included:

Microwave.
Satellite.
Coaxial cable/fiber optics/wire.
Atmospheric optical systems.

There is no standard definition of bypass. Consequently, many "bypass studies" are really not comparable because of inconsistent definitions. To avoid misunderstanding, it should be noted that we did not classify as bypass any communications systems that are not normally employed by local telephone companies. Consequently, we did *not* classify as bypass the following:

Wideband local area networks in a single building.
Inside wiring.
Use of exchange carrier private lines.
Replacement of a Centrex with PBX.
Receive-only video systems used by cable television systems.
Mobile radios connecting base stations with vehicles.
Use of an intra-LATA toll carrier other than an exchange carrier *when the exchange carrier provides access.*

Without bypass, a customer would utilize the local telephone network. Once connected to the network, a customer may originate a call to, or receive a call from, any other subscriber connected to the network regardless of whether the calls are intraexchange, interexchange, intrastate, or interstate.

Local telephone companies require sufficient revenues to cover their costs and make appropriate rates of return on the services they provide by means of their networks. When bypass occurs, large users (such as teleports and their customers) divert substantial traffic from the local telephone company's plant and route the traffic by an alternative local distribution system. Consequently, revenues are reduced, and all remaining customers have to pay higher rates to adjust for the withdrawal of large user traffic from the local network. It is this descending spiral of increasing large-user withdrawal and higher prices that worries the FCC, Congress, telephone companies, and state regulators.

Teleports are generally considered to be potential sources of bypass by local telephone companies. Three key factors associated with bypass seem to be present: (1) large users motivated by lower costs and/or better service, (2) interexchange carrier involvement, and (3) alternative transmission technologies. However, reality is somewhat more complex than this simple threat perception. As discussed below, teleports may be more opportunity than threat. The future remains to be determined.

THE RELATIONSHIPS OF TELEPORTS AND EXCHANGE CARRIERS

Teleports perform a series of functions, each of which has relevance for local exchange carriers. Although there is some uncertainty as to the full range of functions teleports will ultimately perform, three core functions are associated with currently operating and planned teleports: satellite gateways, alternative distribution systems, and real estate development.

Satellite Gateways

The original concept was that teleports would function as centralized gateways to originate and terminate the space segment portion of satellite transmissions. In effect, teleports would be large antenna farms where the satellite traffic of various carriers and private parties would be concentrated, much as air, rail, and truck traffic moves through a limited number of centralized hubs on its way from its point of origin to its destination.

For a local exchange carrier to have such a satellite gateway in its service territory represents an opportunity rather than a threat. A teleport, if it fulfills its expectations, should stimulate both originating and terminating traffic. Since most of that traffic will have to be carried from, or delivered to, the teleport, this provides an opportunity for the local exchange carrier to provide last mile-first mile service.

Afterall, an exchange carrier's basic business is local access and transport. If a teleport increases the total volume of traffic, then the local exchange carrier stands to benefit. This is true regardless of whether dedicated or switched exchange carrier circuits are used to move traffic to and from the teleport.

"Alternative" Distribution Systems

By necessity, teleports need to be located within easy access distance of one or more major metropolitan areas. The issue of critical importance to local exchange carriers is by what means traffic is moved to and from the teleport.

If traffic is transported by a local exchange carrier, the teleport constitutes a business opportunity. However, if a teleport operator constructs (or contracts for) an alternative transport system element to carry traffic to and from the teleport, then the local exchange carrier will be adversely affected, since the local exchange carrier will be "bypassed" by the teleport operator's proprietary system.

The use of nonexchange carrier facilities to move teleport traffic between a metropolitan area and a teleport is the primary bypass threat presented by teleports. The threat is not the teleport itself, but rather who provides access to and from the teleport. For example, the Staten Island Teleport venture of Merrill Lynch, Western Union, and the Port Authority uses a proprietary fiber optics system to move traffic between Staten Island and Manhattan. This represents a loss of revenue to New York Telephone.

Real Estate Development

The third major function associated with teleports is real estate development. Commercial office space is constructed adjacent to a teleport. Theoretically, (1) the teleport will attract tenants to the location, and (2) tenants will pay a premium to locate in space with the high tech image of an adjacent teleport. This concept still remains to be proven in the marketplace.

More often than not, the real estate development near the teleport will involve shared tenant services. The shared tenant services concept can have several revenue effects from the viewpoint of a regulated telephone company:

> To the extent the provider of shared tenant services interconnects directly to an interexchange carrier, local telephone companies lose access charges, as well as possible other local transport charges. (Obviously, if a real estate development is collocated with a teleport, direct access to interexchange carriers without using local telephone company plant is feasible and potentially cost effective.)

If a real estate developer (or an agent for the developer) with multiple properties in an area privately interconnects those properties (e.g., via private microwave or fiber optics), then the local telephone company loses additional local traffic and the associated revenue from that traffic.

The effect of using a large, shared PBX for small and medium tenants will be to reduce the number of telephone company local loops required to serve the tenants in a given rental property.

The first two effects cited above involve actual bypass of telephone company facilities. The last effect is not really bypass but reflects the result of concentration when the traffic of small and medium users is aggregated for the purpose of calculating PBX trunk requirements.

BYPASS AND THE BUSINESS OF TELEPORTS

If teleports can be operated successfully as a business enterprise providing satellite gateway services, then teleports present opportunities to local exchange carriers because of the stimulation effect on traffic transported to and from the teleport. This assumes, of course, that local exchange carriers can successfully compete so that traffic is carried to and from teleports on switched and/or dedicated exchange carrier circuits.

However, if teleports are constructed with proprietary private distribution systems, then the exchange carriers will lose revenues, and the teleport is transformed from opportunity to threat. Furthermore, if real estate developments with shared tenant services become an integral feature of teleports, then this can constitute an additional source of lost revenues to an exchange carrier.

Only after the marketplace sorts out what is the real business of teleports can the full bypass implications to exchange carriers be identified. As long as exchange carriers provide access and transport services to teleports, the exchange carrier will derive a benefit from the existence of the teleport. In fact, there is no reason teleports and exchange carriers can't coexist and be mutually beneficial.

■ APPENDIX ■

Excerpt of testimony of Dr. Joseph S. Kraemer before the Public Service Commission of Wisconsin, 1985.

DEFINITION AND NATURE OF BYPASS

Q. For purposes of your study, how was bypass defined?

A. For purposes of our study, bypass was defined as the origination and/or termination of a call without the use of a local telephone company's plant. Bypass results in a loss of revenues to the local telephone company.

Q. Dr. Kraemer, would you please explain how the revenue loss would occur?

A. Absent bypass, the customer would utilize the local telephone network. Once connected to the network, a customer may originate a call to, or receive a call from, any other subscriber connected to the network regardless of whether the calls are intraexchange, interexchange, intrastate, or interstate.

Local telephone companies require sufficient revenues to cover their costs and make appropriate rates of return on the services they provide by means of their networks. The revenues may come from local exchange services, intra-LATA[2] private line services, access charges, and intra-LATA toll services. *Bypass reduces revenues from one or more of these service categories.*

When bypass occurs, the local telephone company experiences a reduction in revenues from local subscribers, the interexchange carriers, or both. I've prepared a schedule to illustrate how this occurs. Schedule 1a shows a simplified configuration in which a customer has two locations (A and B) in a single LATA. The customer uses the local telephone company's plant to route traffic between locations A and B and between either location and the long-haul transmission facilities[3] of an interexchange carrier.

In Schedule 1b, the customer at location A can now bypass the local plant using dedicated bypass facilities to reach either the interexchange carrier's point of presence or location B. Since the circuits are bidirectional, the traffic routed outside of the local plant in Schedule 1b can originate at customer location A, customer location B, or outside the LATA (entering through the interexchange carrier's point of presence).

It should also be noted that subsequent to January 1, 1984, interexchange carriers can locate within a LATA as many POPs and at whatever location they desire. Since one component of access charges is distance sensitive, more POPs located closer to large customers will reduce the owner's cost of access (and also Winconsin Telephone's access charge revenue). Although not complete bypass, this is a form of partial carrier bypass.

The effect of bypass is a loss of revenue from both local subscribers and interexchange carriers. As shown on Schedule 1b, in a bypass configuration traffic is diverted from the local telephone company. The lost traffic means lost revenue. *Furthermore, the remaining subscribers must pay higher rates in order to compensate for the revenue lost through bypass.*

Q. How could this happen?

A. If bypass occurred, large users (which are generally businesses) would divert substantial traffic from the local plant and route the traffic by another means. The effect on the telephone company's costs would be as follows:

1. Nontraffic-sensitive costs (e.g., depreciation) associated with the plant idled because of bypass would not be reduced.
2. Some traffic-sensitive costs might eventually be reduced. However, operating and maintenance costs would take some time to adjust to the reduced traffic on the network.

The loss of revenue without a proportionate reduction in costs would reduce the company's net operating income. A reduction in net operating income can be expected to adversely affect the telephone company's ability to earn its authorized rate of return and its ability to raise capital. After all, once divestiture occurs, the Bell Operating Companies (BOCs) must compete in the marketplace against one another, AT&T, and other corporations for capital. Erosion of financial performance because of bypass will increase the cost of capital in the marketplace as investors demand larger returns to compensate them for the increased risk of investing in a telephone company experiencing income losses due to bypass.

In addition, all remaining customers (including small business and residential customers) would have to pay higher rates to adjust for the withdrawal of large user traffic from the local network. This would occur because the remaining customers would have to assume the revenue requirement for essentially the same level of telephone company plant and expenses. The effect of concentrating capital recovery charges on a shrinking customer base would encourage additional bypass.

The risk of loss is increased because local telephone company revenues tend to be concentrated disproportionately among a small number of business customers. In addition, revenues tend to be concentrated geographically.

Q. Would you please elaborate?

A. Schedule 2 shows that a large proportion of Wisconsin Telephone's business MTS revenues are generated by a very small percentage of Wisconsin Telephone's business customers.

SCHEDULE 15–1*a*
Local Loop Configuration—No Bypass

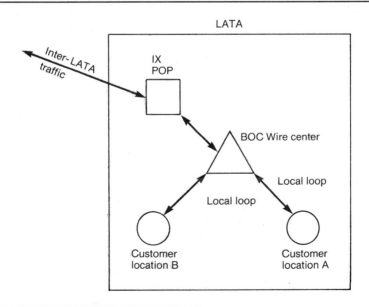

SCHEDULE 15–1*b*
Bypass Configuration

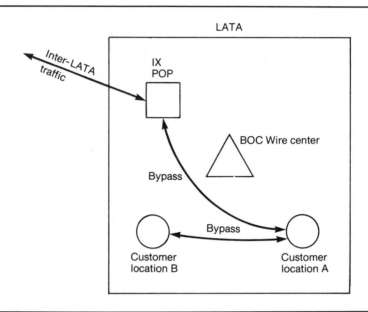

This concentration increases the vulnerability of local telephone companies to bypass since revenues can be disproportionately affected whenever (1) major business customers move to a bypass system and/or (2) bypass technologies are introduced into the small geographic area covered by high-revenue wire centers. This last point on geographic concentration deserves emphasis. It means that *a competitor to the local telephone company need only build a bypass system to handle high-volume traffic originating from a limited number of large customers in a small geographic area. The competitor does not have to replicate the entire ubiquitous local plant.*

■   NOTES   ■

1. To be included within the scope of the Touche Ross study, a user had to spend a minimum of $250,000 annually for exchange and interexchange services in a given state.

2. LATA is an acronym for Local Access and Transport Area; this term refers to the geographic service area in which a Bell Operating Company (BOC) can provide service. A BOC cannot provide inter-LATA service; the carriers who provide such services are designated as "interexchange carriers."

3. The entry point to the long-haul transmission facilities of an interexchange carrier is known as the carrier's point of presence (POP).

Author

Dr. Joseph S. Kraemer
Partner, Touche Ross & Co.
Washington, D.C.

Joseph S. Kraemer is a partner in the Management Consulting Division of Touche Ross & Co. Operating from his office in Washington, D.C., he coordinates a nationwide telecommunications consulting practice. Kraemer has experience in satellite, microwave radio, cable television, cellular radio, and other technologies. He specializes in competitive analysis, market research, new ventures, and communications system selection. Kraemer is a frequent speaker at industry conferences and has testified before numerous regulatory commissions. He is married, has two children, and lives in northern Virginia.

Chapter 16

The Role of Local Cable and Fiber Optic Networks to Corporate Voice and Data Users

Carl Gambello

Manhattan Cable TV

Mel Van Vlack

Manhattan Cable TV

Outline

CABLE COMPANIES AND INSTITUTIONAL NETWORKS (I-NETS)

From a modest beginning with a primary purpose of improving broadcast television reception, cable TV has gone on to provide 20, 36, and then more than 100 channels of TV capacity. Along the way it has added new functions—pay television, public access programming, simple versions of consumer polling, and more. All of these uses are intended to serve, entice, and entertain the home market. Recently, however, the ability of cable TV to serve different markets with different kinds of services has begun to gain attention. Among these new services are those aimed at the business market and provided over an institutional network, or, as it is beginning to be called, an I-Net.

Institutions and organizations—rather than residential subscribers—have become a new market for cable television. The services this market is interested in buying are communications services, such as data transmission, videoconferencing, and private phone system link-ups. To serve this institutional market, cable systems are being built with a coaxial cable that is separate from the wires that bring entertainment and other services into residential homes, "downstream" from a main transmission facility. This separate cable allows users in institutions and organizations to send as well as receive communications. This means that an institutional user can send a signal to one or more other institutional users and receive a signal coming from another user or from the cable system's headend facility. These new nonentertainment uses are known as institutional uses, and the cable plant that is dedicated to serving organizations rather than homes is known as an institutional network.

Institutional networks exist primarily in cable franchise contracts. Relatively few cable companies have yet translated I-Net contract promises into actual operations, and most of these promises, as well as actual builds, are still in an early stage.

Like most innovations, I-Nets did not spring fully grown from the minds of cable or municipal dreamers. Rather they grew in response to institutions, both public and private, that conceived of ways coaxial cable might solve a problem or realize an idea.

The I-Net usually carries anywhere from 20 to 60 television channels. It can integrate audio, video, data, and text services. With the right end-user equipment, the I-Net can also provide voice grade (telephone) service. Unlike the primarily "downstream" residential cable, which in newer franchises will also contain about four upstream channels, the I-Net is generally planned so that there is an approximately even division between upstream and downstream channels. This "mid-band split" thus provides an equal number of channels on which to send and receive signals.

I-NETS AND THE MARKETPLACE

To understand the commercial niche cable companies are trying to carve for themselves, it's important to look at the new telecommunications marketplace and to understand some basic facts about its needs. Telecommunications managers are continually seeking means to transmit information at higher speeds and lower costs. Additionally, they seek to integrate and simplify the transmission of data voice and video communications over one medium and with as few vendors or providers of service as possible.

However, despite technological advances and the emergence of new competitive long-distance telecommunication carriers, the nation's major communication system remains the typical telephone, the 4,000-hertz twisted copper wire pair limited to passing 9,600 bps with special "conditioning." Moreover, telephone switching equipment was planned to handle the simultaneous use of only a certain percentage of telephones in any one area at any one time. It was also designed to accommodate the typical voice telephone user, not one who hooks a computer to the telephone line for a seven- or eight-hour "call."

In contrast, CATV coaxial cable can transmit information at very high frequencies of up to several hundred million hertz. Depending on the cable plant, it can simultaneously transmit up to some 54 different 6 million-hertz channels. Though not a switches system, in contrast to the narrowband telephone, cable is an extremely wide transmission medium able to handle very high transmission rates and the integration of data, voice, and video transmission.

More and more cable plant is being installed in local areas, passing increasing numbers of homes and businesses. By 1983 the country's 50 largest cable operators had installed nearly 400,000 miles of cable plant; and in 1984 they installed another 44,000 plant miles in their local franchise areas. It is the cable company's ability to provide economic high-speed wideband transmission in the local loop over its I-Net or I-Net-like plant that gives cable the opportunity to develop a position for itself in the evolving commercial telecommunications marketplace. It is a position that could grow stronger as increasing numbers of telecommunications users seek to send more wideband signals within the local loop and over long distances into the local loop—that last mile that remains the most difficult and costliest to cover.

THE ROLE OF LOCAL CABLE NETWORKS

Tenants attracted to the high tech building because of improved telecommunications capabilities inherent in its design will want to have the same advantage when communicating out of the building to other locations in the vicinity or to a common carrier for longer distance links. In many cases, a customized local cable network can be used to gain that advantage.

Using presently proven technology, a local cable network can provide reliable high-capacity communications within a radius of at least 10 miles of the high tech building. Fast-paced technological advances, particularly in the field of fiber optics, are increasing that range rapidly. These cable systems can therefore be used to interconnect buildings in a campus environment, to link multiple locations within a city, or to provide connection to common carrier nodes, thereby gaining improved access to intercity and international facilities.

DATA TRANSMISSION AND CABLE I-NETS

Several types of cable can be included in a local cable network, each type having inherent functional characteristics that make it optimal for specific uses. Cable systems can therefore be designed using several types of cable combined in customized configurations in order to meet a diversity of applications. Those applications can include data, voice, and video requirements in a multitude of combinations.

The data communications capabilities of local cable systems include, but are not limited to, the following types of communications links.

Dedicated point-to-point links can be used to interconnect two computers, two terminals, or a computer and a terminal on a permanent basis. Any standard data rate up to and including 1,544,000 bits per second can be accommodated. Equipment for still higher rates is now becoming available. Higher data rates can also be used with end equipment that multiplexes, or combines, data from many devices onto a single data link.

Dedicated point-to-multipoint links operate in a similar way. However, these links provide permanent connection between a computer and a number of terminals or other devices in order to enable the computer to exchange data with each on a sequential basis. Dedicated point-to-point and multipoint links are functionally comparable to traditional services offered by local telephone companies. However, such standard services typically operate only up to 9,600 bits per second. The much higher rates provided by local cable systems are therefore a distinct advantage.

Multipoint switched links allow many computers, terminals, or other devices to be connected to each other dynamically through the use of addressing. Any such device can then communicate to any other device on the link on a temporary, switched basis. This link is a functional extension of the local area networks (LANS) being marketed by a number of companies for office automation purposes. The local cable system can be used to provide multipoint switched links, or LANS, that encompass several buildings. They can also interconnect independent local area networks in several separated buildings and provide access from devices on a LAN to long-distance services from a common carrier. These links are equivalent to standard telephone switched services. They provide higher data rates and, in most cases, greater flexibility and ease of operation.

The high capacity of the cables used in the system can permit many links of the three types described to operate simultaneously on a single cable.

VOICE SERVICES AND CABLE I-NETS

One of the important advantages offered by a high tech building is the availability of improved voice service inherent to the use of a private telephone system. In many cases the building owner or manager provides the necessary wiring from each floor to an advanced telephone switching system, which is located within the building for the shared use of tenants. Such a system offers low-cost interconnection of the telephones within the building and many also include many advanced features designed into newer telephone systems. Typical of such features are teleconferencing, call forwarding, least-cost routing, and detailed call accounting.

Where there is a complex of buildings or where a a particular tenant occupies several buildings in the same area and wants his or her own system, the private telephone switch can be divided and a portion of the system extended to, and included in, each building. It's also possible that each building may require its own independent telephone system. In either case voice links between buildings are necessary.

The local cable network can be used to provide those links. It can also provide required connections between a private telephone system and a common carrier for longer-distance voice communications. The cable system offers several options for providing such local voice services. The primary consideration in the selection of an option is the number of voice links needed.

Small numbers of point-to-point voice links can be transmitted through the cable system on an individual basis. Such an approach is most practical when several end points require roughly 10 or less connections between them. It also provides flexibility when the link end points are subject to relocation.

In cases where larger numbers of links, or trunks, are needed, a local cable system can provide several options. Equipment called channel banks, or their equivalent, can encode and combine several voice signals into a single binary bit stream having a data rate of 1,544,000 bits per second. That signal can then be transmitted through the cable system in a fashion identical to that of the high-speed dedicated point-to-point data link discussed previously. Present standard channel banks can combine 24 voice signals. However, more sophisticated equipment now becoming available can increase this number to 48 signals or more. This technique is practical when fixed-end equipment, such as telephone switches, must be linked by at least 10 voice trunks.

Other types of equipment can combine even greater numbers of voice signals in either encoded or nonencoded form and provide transmission by direct connection to a dedicated cable. For example, some telephone switches are now capable of combining several hundred voice signals to produce a 45 million-bit-per-second binary-encoded stream of light pulses. For transmission, this pulse stream can be coupled directly to a single optical fiber contained in the local cable system.

VIDEO SERVICES AND CABLE I-NETS

The cable system can also satisfy a number of video transmission requirements. Video images can be transmitted through the cable in either an encoded for unencoded form.

Equipment, called a codec, is now available which can sample, encode, and simplify a video signal to produce a digital pulse stream at a standard data rate. Present codec designs permit full-motion video to be transmitted at a rate of 1,544,000 bits per second. To the cable system, that signal appears the same as a high-speed data or encoded voice signal. Freeze-frame video, where the image is changed every few seconds, now requires a 56,000 bit-per-second data link. Development of improved equipment will lower the data rates needed. A codec-equipped videoconferencing room in the high tech building can therefore be connected through the local cable system to conventional long-distance data services.

Standard unencoded video links can also be provided between buildings or to long-distance carriers. A security system encompassing several buildings is a typical application requiring such links. The high-resolution video required by suppliers of programming services and by medical diagnostic systems requires more capacity but can be easily accommodated on the local cable networks.

CABLE TYPES AND CHARACTERISTICS

Several types of cable can be included in a local system designed to satisfy the applications discussed above.

Twisted Pair

Cables composed of twisted copper wire pairs have been used for traditional telecommunications services for many decades. They remain the predominant facility offered by local telephone companies. Each pair can provide a point-to-point connection for a single signal. Although designed primarily for voice purposes, a twisted pair can also be used for data when equipped with modems as end equipment. Typical modems convert data signals to voice form for transmission on the cable pair. The maximum data rate achievable is usually 9,600 bits per second. Other modem types can transmit data signals directly over the cable for a maximum distance of several miles. The distance of possible transmission decreases proportionally as the data rate increases. The maximum data rate achievable between buildings with this approach is typically 56,000 bits per second. For this reason, copper pairs are not suitable for most high-speed data or for video applications.

A wire pair is needed for each signal; therefore, cables of this type are quite large and bulky compared with newer cable types having greater capacity. Twisted pair cables are also more susceptible to a number of electrical noise and interference conditions, which can disrupt proper operation. They are relatively easy to tap for signal monitoring purposes and therefore provide less security for information of a private nature. For these reasons, twisted pair cable has relatively limited usefulness in most local cable networks.

Coaxial Cable

Coaxial cable was developed to meet new requirements beyond the capabilities of twisted pairs. Applications with such requirements included radio, television, radar, and video signal transmission. The telephone industry also began to use coaxial cable for the bulk transmission of many voice links combined into a single signal.

The physical structure of coax makes it suitable for such applications. It transmits its signal energy through a single center conductor, which is totally surrounded at a uniform distance by a shield and ground return, which serves the same function as the second wire of a twisted pair. This structure permits coaxial cable to have a signal capacity many times that of twisted pair and also provides effective shielding against electrical interference. In recent years the improved performance of coaxial cable has caused its use to be expanded to include a variety of new data and voice applications.

Baseband. These applications can be functionally divided into two groupings: baseband and broadband. In systems using baseband operation, a single digital data signal, having a rate up to 10 million bits per second or greater, is transmitted on the cable at any instant of time. It can be sent several thousand feet without having to be regenerated to its original form. In many operations, interface equipment is located at either end of a point-to-point cable. As an alternative, multiple signal sources may be easily tapped onto the cable throughout its length in order to share use of the cable capacity on a time sequential basis. A local area network, or LAN, is an example of such an application. A typical LAN permits many compatible interface devices to communicate with each other on a dynamic, switched basis. The high cable data rate permits each system user to operate as if a private, dedicated connection was being provided. A local cable system which includes coax can provide the baseband applications described above within a complex of adjacent high tech buildings.

Broadband. Broadband coaxial cable technology has been largely developed to satisfy the need of the cable television industry to distribute a number of high-capacity video signals throughout a city-sized area on a single cable and to tap those signals into a great many locations. This is accomplished by assigning each signal a separate portion of the overall cable capacity, that portion being proportional to the information content of its assigned signal. Since total coax capacity can be as much as 100,000 times that of a conventional voice twisted pair, a great quantity and variety of signals can potentially be accommodated. Present techniques permit at least 50 unencoded television signals on a single cable.

In recent years, techniques have been refined to permit two-way data and voice also to share the active cable. Present equipment permits transmission of about 130 two-way data signals of 1,544,000 bits per second. Alternately, several thousand lower speed data or voice connections can be provided. A flexible mix of data, voice, or video signals is possible.

Amplifiers spaced every 1,000 to 2,000 feet permit distribution of high-quality signals throughout a radius of 5 to 10 miles. Beyond that distance quality begins to gradually decrease. Within its area of operation, a broadband cable system can be easily tapped to provide use of the cable to many buildings and to many locations within each building. A well-designed broadband system will provide preinstalled access to the cable at all locations having potential need for its use. New applications can therefore be readily connected into the cable and can then communicate with any other access point, or combination of access points, on the system. Several suppliers are now providing local area networks that are compatible with broadband operation. Using this approach a LAN composed of several hundred terminals can be accommodated on a broadband cable in the capacity normally allocated to a single television channel.

A local cable system that includes broadband coaxial cable will provide its users with great capacity and flexibility in communicating within high tech buildings and to the outside world.

Optical Fiber

The development of fiber optic systems has become a very rapidly evolving part of the communications industry. This extraordinary growth is based on relatively recent breakthroughs in several interrelated areas of technology.

Strands of ultrapure glass fiber are capable of transmitting light signals for long distances. Those light signals can be varied in several ways in order to convey intelligence. The technique most applicable to telecommunications functions involves transmission of narrow pulses of light coded to represent voice or data information. These light pulses are produced by electronic components called laser diodes in response to comparable pulses of electrical current. Light coupled onto the fiber and transmitted through it is sensed at the far end by complementary components, which convert the light pulses back to electrical form.

Rapid advances in the technology of both fiber and end equipment are serving to enhance several advantages inherent in the use of optical fiber systems. Improved end equipment permits use of progressively narrower pulses. New fiber types can transmit those pulses with acceptable distortion. As the pulses become narrower, more of them can be transmitted in a given time; and therefore, the capacity of the system is increased. Advances in the capabilities of end equipment, in the purity of glass used in producing fiber, and in the techniques for splicing sections of it together result in less attenuation of the light. This permits communication over increasing distances without the need for pulse regeneration. Many fiber systems now in operation can transmit approximately 100 million pulses per second for distances of 5 miles or more. A newer fiber type, now readily available, increases both speed and distance by a factor of at least two. In a number of labs, speeds of over a billion pulses per second and distances approaching 100 miles are being achieved.

The failure rate of end equipment components has been greatly reduced in recent years. Coupled with the stability of the glass fibers themselves, this results in highly reliable operation. This reliability is enhanced by the fact that considerable distances can be achieved without the use of intermediate equipment.

Optical fibers are approximately the diameter of a human hair. Newer types are considerably finer. Many fibers, having a tremendous collective capacity, can therefore be included in a composite cable of reasonable size. Individual fibers can then be broken out of the cable at strategic points to serve specific functions.

Because only light energy is transmitted through a fiber system, it is immune to electrical interference of all types. This is particularly important when high-quality transmission is required in the vicinity of electrical power facilities, near an elevator system or any other equipment requiring surges of power.

Its physical characteristics and mode of operation make fiber a very secure transmission medium. It is relatively difficult to remove the surrounding protective material and then the opaque coating of a fiber in order to gain access to the information bearing pulses of light. It is probable that the effort to do so will disrupt proper operation and therefore be detected. Highly sophisticated equipment is also needed to detect the light signals and to interpret their informational content.

The same factors that make fiber secure also make it relatively difficult to work with. Coupling light between end equipment and the fiber, splicing the fiber, and testing its operation are all rather precise operations requiring complex equipment and a high level of training.

At the present time most fiber optic systems are configured similarly to those using twisted pair cable. Each fiber conveys a single signal between two fixed end points. Equipment to divide the capacity of a fiber to permit simultaneous multiple signals is now in laboratory development. Improved techniques to connectorize, splice, and tap fiber to multiple locations are also being developed. It will probably be several years before these developments result in practical and inexpensive products suitable for general use.

Coaxial Cable versus Optical Fiber

For the foreseeable future, coax and fiber will each continue to have advantages relative to the other. Fiber provides reliable, high-capacity, point-to-point links over relatively long distances. Coax, on the other hand, can handle many independent signals simultaneously and distribute them among many end points in a very flexible and timely fashion. Many local cable systems will therefore contain both types of cable, each being used in the applications for which it is best suited.

There are several potential sources of local communications capability. In many cases, it may prove advantageous to use a combination of such sources in order to develop an overall system.

Local telephone companies are working hard to upgrade their cable plant facilities. They are removing voice-oriented equipment from a portion of their twisted pair cables in order to permit customers to transmit 56,000 bps data, or higher, to the nearest central office. They are also installing fiber optic capabilities between central offices and to the locations of prime customers having high-speed data requirements of 1,544,000 bits per second or more. This is a major effort, and some customers may encounter significant delays in receiving such services for several years to come.

Leasing the high-speed telephone company services described makes the most sense when the services are used to provide access to comparable capabilities from long-distance carriers or to interconnect high tech buildings beyond the range of a local cable network.

In some instances, telephone companies have installed dedicated cables for specific customers and have leased those cables on a long-term basis. However, provision of such services is not a primary business for most telephone companies. Therefore, they are not expected to be a major supplier of dedicated cables for inclusion in local distribution systems.

Cable companies supplying CATV services have access to cable distribution facilities within their franchise area. They also can have new facilities constructed if they are needed. This is an important advantage, particularly in urban locations where ducts in the underground system are often scarce, and a considerable level of expertise is required to use them advantageously.

This local distribution capability makes the cable company a natural partner or supplier of services to its neighboring teleport. The teleport provides a long-distance access point and customer facilities at that location. It may also extend a backbone network to several node points in the urban area served. However, access from those nodes to a large number of off-site customers for a variety of services requires creation of an extensive local communications network. This is typically an expensive and time-consuming task not included in the primary charter of the teleport. The cable company may already have such a network in place. If not, it will probably have the capability to develop such a network in a timely and efficient manner. This point will be reiterated in the following section, which describes Manhattan Cable Television's data services.

URBAN CABLE TELEVISION SERVICES

Manhattan Cable Television has been supplying a range of services to the New York City business community for a number of years. It will be used as a specific example of the types of capability a cable company can offer.

Manhattan Cable operates a broadband coaxial cable system dedicated to the provision of two-way data communications services. That network is active in more than 70 buildings in New York's prime business and financial districts. In many of those buildings, it is functional on every floor. A number of corporate clients each lease a portion of the system's capacity in order to communicate between buildings at data rates ranging from 1,200 to 1,544,000 bits per second.

For a fixed monthly fee, which is proportional to the data rate used, but independent of a client's particular locations, Manhattan supplies and maintains all facilities up to the standard digital interface that connects to the customer's data equipment at each end of the link. Maintenance service is available around the clock, 7 days a week, with a typical response time of less than two hours.

In addition to monthly costs, clients pay a nominal installation charge. When services are required in an unwired location, users share in costs to bring the cable into the building and to make it active. This tends to limit the customer base to companies having substantial communications requirements.

System Usage

Clients are attracted to use the system for a variety of reasons. Some require the quality operation and high data rates provided. To others, availability is of great importance. Within wired buildings, new circuits can be added, and speeds can be changed or end points relocated within a matter of days. Services in unwired buildings have a lead time of about three months. The availability of long-term, fixed-rate contracts is also of importance to many customers.

Within the past two years, typical usage of the system has been shifting. The trend has been toward use of higher-speed circuits. Data, voice, and video applications are now all being supported by those circuits. A strong recent trend has been toward greater use of the system to link end-user locations and carrier facilities, usually at high speed.

Use of the system to connect off-site users to the teleport is also in accordance with this trend. The broadband network present in more than 70 large commercial buildings represents an excellent alternative for quick and inexpensive access to a large customer base. Connection of that network into the teleport node immediately makes all tenants in those buildings available as potential users of teleport services. The range and quality of capabilities provided by the MCTV network also corresponds to typical teleport specifications. Other cable companies have not developed capability comparable to that of MCTV as yet. However, several have smaller functional networks in place, and a number have developed I-Nets that could be readily adapted for use as local access to teleport facilities.

The success of its shared data system has caused Manhattan Cable to initiate several steps to expand and augment its operation. New equipment is being implemented to functionally double the capacity of the system. Redundant facilities are being supplied for the primary cable network. Switchover capability is included, which will permit immediate restoration of service. A diagnostic system will sequentially test operating parameters of all active system components and report any abnormalities. Through its use, a source of failure will be immediately pinpointed, and marginal conditions can be reported and corrected before they result in an actual outage. Manhattan has no immediate plans to expand its services to include responsibility for such end equipment as terminals, data multiplexers, and channel banks.

Several users of the shared network have also contracted for similar private-use systems between a number of buildings they either own or occupy as a prime tenant. In such cases the client pays for the construction and activation of the system under a long-term contract. Manhattan Cable again supplies all cable equipment and maintenance for which the client pays a fixed monthly fee per circuit. However, a discount from normal rates is provided in order to compensate the customer for initial implementation costs.

A dedicated system can be custom designed to meet the particular needs of the client. Techniques such as the use of a ring configuration and alternate cable routing are typically used to provide a highly reliable system. Provision of online spares for all active components is also possible if required.

Installation and maintenance of dedicated cable links, is now becoming a major segment of Manhattan Cable's nonentertainment business. Its access to the underground duct system and its experience in the use of that system permits Manhattan to install and maintain private cable systems customized to meet the particular needs of the customer.

Customer Uses

Several companies have installed their own cable systems within a number of buildings. The interbuilding cables supplied by Manhattan serve the purpose of linking those systems together into an overall network. Other customers need high-capacity links to tie together telephone switch equipment located in dispersed buildings. Carriers having several node sites within the city need to interconnect those nodes. Either coax or fiber cable can be provided to serve those purposes. In some cases, it is farsighted to install both types of cable at the same time.

Connection of off-site customers to teleport facilities is another practical use for a dedicated cable link. The need for such a link, as opposed to use of the broadband system described above, can be dictated by a number of factors. Included among them are close proximity of physical locations and large building transmission requirements or special nonstandard applications. By the nature of their business, cable companies in general can be an excellent source of dedicated cable facilities.

The customer typically pays the initial cost of cable installation and test. Some clients also choose to supply the cable used. Manhattan assumes responsibility for the installation and then maintains the cable under a long-term contract. The monthly maintenance fee is based upon the number and types of cables involved and upon the length of those cables. Periodic tests and preventive maintenance of the cable system can also be performed at specified intervals for an additional charge. Thus far the users have assumed responsibility for all end equipment connected to the cables.

Based upon the experience of Manhattan Cable and on the growing demand for local cable capabilities, a growing number of cable companies are now beginning to offer services such as those described above.

Some companies desire to own all or part of their local cable network. They therefore assume full responsibility for its design, installation, and maintenance. This option is most practical in a controlled environment, such as a college campus or hospital complex, where cable installation and maintenance do not become a major problem.

Cable Installation

In an urban location, cable installation and maintenance presents much greater difficulty. Access to the duct system is rigidly controlled. In New York City, private companies may gain limited access to available ducts for their private needs by making application to the proper city agency. This can be a complicated and lengthy process. In many other cities, this is not possible at all as yet. Maintenance is also a primary consideration.

A very limited number of companies are permitted access to the underground system in order to test and repair cables needing maintenance.

A limited number of firms are now offering consulting services geared toward gaining of access to the duct system, design, installation, and test of cable networks and provision of cable maintenance. Services from such firms naturally add to the cost of establishing a private cable network. The expertise needed to perform the services mentioned above is now limited to a relatively few individuals. It is therefore difficult to engage a firm that can supply the full range of services and do a turnkey development of a private cable system. It is more common to hire several firms, each contributing to a particular aspect of the overall job.

Technical Advantages

Independent of their source and ownership, important technical and operational advantages result from the use of improved types of coaxial and fiber optic cable. The great capacity inherent in their design also produces several related improvements. The bit error rate performance of links using these cables is superior to that provided by the use of twisted pairs. Better immunity to noise and interference also contributes to this improved signaling capability. The relatively small number of cables required makes it much simpler to install redundancy and switchover, with greater reliability being the reward. Their limited number also makes it much easier to implement diagnostic and test capabilities on the cable and quicker to repair cables in the event of a failure. Speed in the modification and initiation of applications is greatly improved because of the preinstalled capacity provided by a well-designed cable network.

Financial Options

As discussed above, there are a number of options available when implementing a local cable capability. The traditional approach has been to lease all facilities and to pay on a monthly basis. Fear of rapidly escalating costs and the availability of viable alternatives has led to the present trend toward more independence and innovation. Users are now more willing to assume significant operational control and responsibility relative to their communications facilities, and many will make significant capital investments in equipment and cabling in order to reduce or guarantee long-term expense. A preference for longer-term contracts is also becoming more common. The benefits of this strategy can be substantial.

Several of the projects in which Manhattan Cable has participated resulted in payback periods of one to two years. The savings in the longer term will prove to be very substantial. In addition to these direct financial benefits, assumption of greater control, if effectively exercised, can result in improved operation and faster maintenance response by personnel who are more directly accountable to management.

This more aggressive approach on the part of communications users is an integral part of a very rapid evolution taking place in the entire communications industry. The breakup of AT&T, coupled with a general trend toward deregulation, has provoked an era of unprecedented competition. A great number of new carriers are offering improved transmission options. Rapid technological progress, particularly in the field of microelectronics, has produced a flood of new hardware and software products offering greatly expanded capabilities. Communications users are rushing to develop the expertise needed to function effectively in this new era. Consultants are gearing up in order to provide proper help where needed.

The primary focus of much of this activity is now shifting from long-distance communications to the local connection. There are great savings and improvements to be made in local communications and in gaining efficient access to the long-distance network. The concept of the high tech building lies at the heart of an effective local communications strategy. Functional, efficient, and reliable local cable networks will provide the webs of wire and glass needed to tie it all together.

BENEFITS TO THE I-NET USER

Why would a business buy telecommunications services from a cable operator? What are the benefits of such a service? How cost effective is it?

In a few cities today, cable offers business wideband, high-speed capacity that may be: (1) unavailable elsewhere, (2) available only at greater cost, or (3) obtainable only with much greater delay. Currently, just a handful of cable systems serve a handful of commercial users. But two major catalysts could lead business to a greater use of cable services—the divestiture of AT&T and the growing desire of companies to integrate their voice, data, and (in the future) video telecommunications services.

EFFECTS OF AT&T DIVESTITURE

The major impetus to the use of alternative technologies and new vendors is the economic uncertainty caused by the breakup of AT&T. On January 1, 1984, the 22 Bell Operating Companies (BOCs) became officially separated from AT&T.

AT&T has aggregated its remaining units into two major sectors: AT&T Communications and AT&T Technologies. The first consists entirely of the former long-lines division. Under AT&T Technologies are AT&T Technology Systems (formerly Western Electric); AT&T Bell Labs (research and development); AT&T Network Services (selling transmission, switching, and central office products and services to telephone companies and internally); AT&T

Consumer Products (dealing in the consumer market); AT&T International (serving markets abroad); and the AT&T Informations Systems Group. Although the last falls under AT&T Technologies, it is now set up as a financially independent subsidiary that will service specific markets, selling digital PBXs and other business communication products.

As a result of divestiture, AT&T will no longer be a one-stop shopping center for service, equipment, and installation. Also, users will be confronted with new rate structures as they use service within and between the approximately 165 Local Access and Transport Areas (LATAs) that replace the 22 BOC service areas (which have generally followed state boundaries).

Where once service between two cities fell under one pricing pattern, service between the same two cities may now trigger a totally different rate structure depending on whether they are in the same or different LATA. All intra-LATA service will be provided by the respective BOCs, which retain their separate identities within each regional network.

As the United States moves from the dominance of a quasi-national entity, AT&T, to a highly competitive market with many different and unintegrated service and equipment providers, there will be new opportunities for entrepreneurs and users, as well as formidable problems and the potential for chaos.

INTEGRATION OF SERVICES

Data information is no longer the exclusive domain of the data processor and the mainframe computer. Using microcomputers, more and more end users are aggregating, accessing, and manipulating information themselves. Not only do these end users want to use information, but they want, if necessary, to send and receive it within and between buildings and within and between cities.

As greater numbers of people become sophisticated computer and information users, they will want more sophisticated communications services. Also, as the use of telecommunications of all kinds proliferates, the end user, especially telecommunications managers in larger companies, are becoming advocates for the integration of voice, data, and video services.

Integrating services means eliminating or reducing the time it takes to deal with multiple vendors, learn their pecularities, pay multiple bills, and match up and interface multiple kinds of equipment. So when business evaluates the benefits of alternative technologies, it will consider not only cost, but the capacity to integrate services and possibly obtain them from a single vendor, whether a phone entity, a cable company, or a carrier like MCI.

CHALLENGES TO INSTITUTIONAL NETWORKS

Challenges to I-Net expansion come from regulatory bodies and competing technologies. In the technological arena, the cable industry's openness to new technologies and system architecture will determine how competitive cable will be in the next five years. On the legal front, cable operators are active in the courts and before legislative and regulatory bodies, trying to make sure that they do not wake up to find themselves legislated or regulated out of new, nonentertainment businesses. A formidable competitor is the telephone industry, which believes that tariff and regulatory decisions—not technology—will determine the outcome of competition.

CABLE'S UNIQUE POSITION

Today, cable operators are in an excellent competitive position. Laying miles and miles of new cable monthly, they are setting in place a special local loop and carving a niche for themselves in a flexible and growing telecommunications market.

Through I-Nets, cable companies have an opportunity to add new profit centers to their traditional residential base. Commercial users have the opportunity to obtain services that might otherwise be unavailable or available only at a higher price or with greater difficulty.

To enter the commercial market, cable companies will need staffs with particular kinds of technical skills and knowledge, as well as the ability to understand business needs and to tailor cable services to company requirements. The compensation for the effort and investment is a broadening of cable's economic base, new service offerings, and revenue from large-channel capacity.

The small but growing use of cable for business applications is evidence that these offerings have a market. The key for the business user and the cable company provider is getting to know each other and understanding what each can offer the other. If cable can give business what it needs, when it needs it, and at a competitive price, then cable can develop a market position within the continuously expanding telecommunications market. Although important regulatory issues have yet to be settled, the regulatory climate is such that the odds may be in cable's favor.

Authors*

Carl Gambello
Manhattan Cable TV
New York, New York

Carl Gambello's responsibilities at Manhattan Cable have been to effect managerial and system changes within various parts of the organization, including marketing, sales, and operations. His current priorities include directing Manhattan's efforts at the business community, specifically establishing Manhattan's presence as a supplier of data transmission services.

* We wish to express our thanks to Sylvia Hack of Media Enterprises, Inc. for allowing us to use material from the publication, *Institutional Networks: Using Cable for Data, Voice and Video Communications* (New York: Media Enterprises, Inc., 1984).

Mel Van Vlack
Manhattan Cable TV
New York, New York

Mel Van Vlack, prior to his work at Manhattan, was director of communications planning at SIAC. At Manhattan his 20 years experience in teleprocessing, data communications, and RF transmission has enabled him to direct efforts at data facilities upgrade and expansion, introduce enhanced hardware and systems, and develop new market offerings.

Chapter 17

Metropolitan Area Digital Networks

G. William Ruhl
Bell of Pennsylvania

Randall C. Frantz
Bell Atlantic

Outline

Source: Reprinted from Alan D. Sugarman, Andrew D. Lipman, and Robert F. Cushman, eds., *High Tech Real Estate* (Homewood, Ill.: Dow Jones-Irwin, 1985), © Dow Jones-Irwin, 1985.

Central to the continuing advancement of any technical society is the need for an effective communication system. For 100 years the voice telephone met our informational needs. However with the growing use of computers, requirements changed, and demands on informational systems increased. Today's modern, technological society needs more than just voice from its communication systems. It needs new capabilities and services, such as voice with enhanced features, data transfer, and videoconferencing. The growing desire for these and other services is the driving force behind the evolution toward a total service network.

The heart of this evolving communication network is centered in the developing metropolitan areas. Metropolitan areas have always been at the forefront of technological change and innovation. In communications, the metropolitan area will be the focal point of developing an Integrated Services Digital Network (ISDN). This digital network will meet all the varied informational needs of many different users. Customer services will be provided through a single interface to a large integrated network. This network will feature end-to-end digital connectivity and complete customer control of the data transfer process.

To appreciate fully the development toward an ISDN and the contributions of metropolitan area digital networks, a fuller understanding of the information transfer process is needed. This chapter is divided into four sections. The first explains in detail the communication/information transfer process of today's network. The second shows how the network is evolving toward an ISDN. The third section explains divestiture's impact on the network. Finally, we will examine teleports/telehubs and how Philadelphia is a typical developing metropolitan area network that promotes the teleport concept.

COMMUNICATIONS—UNDERSTANDING THE NETWORK

Communications is the transfer of information from one place to another with speed, reliability, and accuracy. Another desired attribute is to achieve this at a reasonable price to the users. The most effective method of accomplishing these goals is through a communications network that is a system of lines, channels, or circuits providing transmission of information. One of the most extensive and sophisticated communication networks today is the public switched network in the United States.

Components of the Public Switched Network

There are three major components to the public switched telephone network: subscriber loop (customer lines), central office (switching facilities), and trunks (lines connecting central offices). Each of these three parts is integrated into one network, which gives the customer the capability to send information from his or her location to any other location where access to the network exists.

Local loop. The subscriber loop provides the customer access to the network. The loop forms the transmission path between the customer's premises equipment and the network over which information can be sent. The customer premise equipment can be a phone, switching system (PBX), data terminal, or video/facsimile equipment, depending on what type of information is being sent. Traditionally the subscriber loop or line was thought of as a twisted copper wire pair, but today other transmission mediums are also used. For high-speed data or video signals, fiber optics provide interference-free, high-quality communications. The copper wire pair, however, is still the preferred choice in many instances for short-distance voice conversation and low- to medium-speed data, due to its ease of installation and low cost. Whether the subscriber loop is made of copper or some other material, the main function is the same: to give a medium for sending information from the customer's location to the next part of the network, the central office.

Central office. The central office is the center of the local exchange carrier's communication network. Through the use of switches and circuits, it establishes a transmission path for sending the information to its final destination. The best way to understand the importance and function of the central office is to consider what the network would look like without the ability to switch calls.

Assume there is a community where eight people live, and each one has a phone. A person wanting to call any of the other seven would need a direct line from one telephone to each of the other telephones in the community. Each of these phones would in turn need separate connecting lines to the remaining ones. A total of 28 lines would be needed to interconnect the eight phones in the community. The number of lines needed in this type of network can be calculated by this formula:

$$\text{Number of lines} = N(N-1)/2$$

where N represents the number of phones in the system.

Such a network would require millions of lines for just a few thousand telephones. This would neither be practical nor affordable.

To improve the efficiency of the network, a switching system located in a central office can be used to receive calls and switch them from one line to another. In this type of network, only one line is needed to connect each phone to the central office. The switch greatly improves the efficiency and lowers the overall cost of the network.

In the early development of the network, switching was performed manually at the central office by an operator. One gave to the operator the name of the party being called. The operator then routed the call to the proper destination by connecting wires on a switchboard. As technology advanced, the method of completing calls became more sophisticated. Direct dial used electromechanical devices, which routed the calls by sensing the numbers dialed by the user as electric pulses. Each set of numbers gave a sequence of commands to a switch, which decoded into a specific location and phone in the network. Today, highly sophisticated computers connect thousands of calls in a fraction of a second. Calls can even be dialed directly overseas.

A call that originates and terminates within the same central office is an intraoffice call. However, if the destination of the call lies outside the area served by the local central office, the call must go over a circuit called a trunk. At the other end of this trunk is the destination central office, which switches the call to the proper subscriber loop. This type of call, which uses two different central offices for completing a connection, is called an interoffice call.

Just as it would be impractical to connect all the telephones in the network with separate lines, it is impractical to have a direct line or trunk connecting all the central offices together. To avoid this, a system of tandem switching centers was developed. A tandem center is a switching office between two central offices that do not have a direct connection. When a central office cannot find an available trunk to the destination central office, it sends the call to a tandem switching facility. This tandem must then decide on how to complete the call. If there is a connecting trunk, it can send the call directly to the destination central office. When there is no direct connection, the call will be sent to another tandem switch. The call can be handled by several tandem switches before it reaches its destination because there are hierarchical levels of tandem switches defined in the network that permit the systematic and efficient routing of calls.

There were four levels of tandem offices in the network prior to divestiture. Now the local telephone companies primarily use the local central office and one or two levels of tandem centers to complete calls. Calls that need a higher level of switching, international or interstate long distance, are handled by a long-distance carrier.

The ability of the tandem offices to search for available routes gives the network its flexibility. If all the direct circuits between two central offices are busy, the switching system can provide an alternate path, called indirect routing. Suppose a call was being placed between downtown Chicago and one of its northern suburbs. The most direct route would be a direct line between the two areas. If direct routing were the only way to complete the call, and if all the trunks were busy, the caller would receive a busy signal and would have to try the call again later. Fortunately, the hierarchical structure of tandem switches gives the system other alternatives. A less direct routing of the call can be made. The call might go from downtown Chicago west and then terminate in the northern suburbs. The system constantly monitors the usage of the circuits, and if they are all busy in one section, the call can be handled by excess capacity elsewhere. Such a process is automatically executed by the system without any user effort. The user is never aware that this process is taking place but thinks the call goes directly from downtown to the north and is only billed for a direct call.

The trunk. We have already briefly mentioned the third part of the network. It consists of the connections between central offices and tandem centers, called trunks. These are higher-capacity facilities, which can simultaneously carry many voice and data channels between offices. Trunks use copper cable, coaxial cable, lightguide or microwave radio for these transmissions. All these mediums are suitable for carrying high volumes of traffic. The backbone of the long-distance traffic was formally carried by cable and microwave radio, but lightguide is now becoming an increasingly popular replacement. This especially holds true as the network converts to the use of more digital facilities.

When information is transmitted over the network, it can exist in one of two forms—analog or digital. An analog signal has certain characteristics that make it very different from a digital signal. Analog is continuous and has an infinite range of amplitudes and frequencies. The human voice is a good example of an acoustical analog signal. When the human voice is converted into an electrical signal by means of the telephone, an electrical analog signal is created that duplicates the acoustical signal. Analog signals suffer from loss, noise, and distortion during the transmission process. To counteract this loss, which is created by attenuation, the signal strength is boosted by amplifiers. Unfortunately, noise (any unwanted signal) becomes mixed with the informational signal during the transmission process. Amplifiers cannot distinguish between noise and the informational signal and therefore amplify both the noise and the information signal. After many amplifications, the signal becomes distorted. Filters help reduce the noise and distortion, but over long distances the signal still loses some quality and part of its informational content. Transmission of analog signals are therefore distance limited by the number of times the signal can be amplified.

Distortion and attenuation of analog signals were tolerable when voice conversation was the only type of signal transmitted over the network. Even if the signal became distorted, the other party might understand enough of the conversation for communication to occur. However, computers are using the network for transfer of data in ever increasing numbers and are not nearly as tolerant of noise and distortion as people are. These problems when dealing with computers would result in data transmission errors. The use of digital signals greatly reduces the potential for transmission errors.

Digital signals differ from analog signals in two ways. Digital signals are discrete in time (noncontinuous) and have a discrete number, not an infinite range, of amplitudes or values. They carry encoded information in a series of on or off pulses called bits. The standard format used in the United States for digital signals is the T-carrier system. The system uses a 1.5 megabits (1.5 million bits per second) transmission rate and can carry 24 simultaneous voice or data channels.

To convert an analog signal to a digital signal is a three-step process: sampling, quantizing, and encoding. First, the signal is sampled at a rate of 8,000 times per second. Each of these samples is then quantized into one of 256 discrete levels by comparing it with a segmented scale. Finally the signal is encoded into a binary word consisting of eight bits. Each of these eight-bit words describes a certain portion of the original analog signal. Using these words, the entire analog signal can be reconstructed at the receiving end of the circuit by reversing the process.

Digital signals are still plagued by the loss, noise, and distortion affecting analog signals, but there is a much more effective way to deal with these problems. While analog signals are amplified to counteract loss of signal strength, digital signals are not. Instead digital signals are completely regenerated at regenerator stations. It is a relatively simple task to build electronics to sense digital pulses and regenerate them. Since the digital signal is completely regenerated or renewed, there are neither amplified noises nor distortions to worry about over long transmission distances. Using this method the signal can be regenerated many times using multiple regenerators. Additionally the chances of inducing errors in the signal are greatly reduced.

Digital signals have other advantages, too. A digital signal carrying voice would require 64 kilobits per second, which is composed of 56 kilobits of voice information and also 8 kilobits used by the network for signaling and supervision of the call. By interweaving signals (multiplexing), much higher bit rates can be achieved, and many signals can be sent simultaneously over the same facilities. For example, 24 voice channels can be multiplexed to produce the T-carrier signal of 1.5 megabits per second and sent over copper transmission facilities. Even higher bit rates, such as 560 Mbps, with the capability of carrying 9,264 voice channels can be achieved using lightguide over only two fiber optic strands. Rates of more than 1 billion bits per second are expected to be possible in the near future. By use of multiplexing, digital signals can achieve even greater use of these transmission facilities.

Although there are many advantages to a totally digital network, it is impractical and not really necessary to convert all the analog facilities to digital overnight. The public switched network was originally designed to handle analog voice conversation, and although analog and digital signals cannot normally be handled by the same equipment, many of the existing facilities have been adapted to handle both types of traffic. This makes possible some services that offer end-to-end digital connectivity.

THE EVOLVING NETWORK

Our present switched network is evolving toward a fully integrated services digital network (ISDN). Although today's network is a combination of analog and digital transmission and switching equipment, the ISDN will be an exclusively digital network. It will support a wide range of voice, data, and video services. The customer will have access to these services through a set of standard interfaces. The first trials of an ISDN are already planned using the international standard "2B + D" channel structure. The 2B + D structure offers two simultaneous 64 kilobit voice/data channels and one 16 kilobit signaling channel. Universal access to an ISDN will not be immediately available, but there is already a wide range of digital and nondigital services available through the network.

The two basic types of informational transmission services provided by the network are switched and nonswitched. Three services best illustrate the difference between these two: WATS/800 service, which is switched; Private Line, which is nonswitched; and Foreign Exchange, which has switched and nonswitched components.

WATS Service

Wide Area Telecommunication Service (WATS) is probably one of the most familiar services available through the network. It permits the customer to make a bulk purchase of long-distance, voice or data service rather than being billed on an individual call basis. This service is very attractive to businesses that do a lot of long-distance calling. The service can be purchased on an outward WATS (bulk charge for outgoing calls), inward WATS/800 service (bulk charge for all incoming calls) or both. No matter which option is chosen, the business has full access to the entire switched network. This service is really quite similar to residential or regular business lines and only differs in the method of billing.

Private Line

A more restrictive type of service is a Private Line (PL). A PL is a nonswitched or hard-wired service. It is very useful for businesses that require immediate or continuous communications with another location. One phone or customer premise switch is connected directly to a second one with no switching facilities. With a PL the other phone can be located anywhere— a different building or even another city. The business customer purchases the exclusive rights to the circuit linking the main phone with an extension. The connection exists constantly and there is no worry about dialing or about the line being busy. This service might be used by brokerage houses to contact the trading floor where immediate, continuous communication is needed. Another use might be a connection between a main corporate computer with branch office computers in other cities or states.

Foreign Exchange Service

A third type of service, Foreign Exchange (FX), contains elements of a nonswitched and switched service. FX is used when a company is located in an area served by one local exchange or central office but wants a phone number in a different central office. An example would be a firm located in Baltimore that wants a local number listed in Washington, D.C. In this case the company would have a normal subscriber line to the Baltimore central office; but instead of connecting to the local switch, it would connect to a dedicated circuit leading to a central office in Washington, D.C. The company would receive its service and dial tone from the Washington central office and not the office in Baltimore. This arrangement is very useful if the company does a lot of business in the area served by the distant central office and wants a local listing in that office's directory.

WATS, PL, and FX give a very good example of the difference between switched and nonswitched services. However, they do not begin to tap the vast potential of voice and data services available through the network.

Centrex

Centrex uses the latest capabilities of the local telephone central office to provide sophisticated communications to the business customer. The central office has a separate line extending out to each subscriber's station. These stations can be telephones, data terminals, or printers. Over these lines the customer can transmit voice or data. These lines differ from residential lines because of the features available.

These features include call forwarding, call waiting, and four-digit dialing to any other station in the centrex. In addition the user has complete access to the public switched network. Centrex Station Rearrangement (CSR) permits the customer to rearrange numbers and features via terminal access to the centrex switch. Centrex also provides digital data capabilities in the central office switch for 9.6 Kbps to 56 Kbps.

There are several advantages to the use of centrex by businesses. As a part of the switched network, service is provided 24 hours per day, even during power outages. There is no bulky equipment located on the customer's premises. There is growth capability both in the number of stations and in the features available. A subscriber can add both lines and features as needed. Billing is provided on a per-phone basis, making accountability and usage recording simple. Finally, since centrex is a leased service, there are no large capital outlays for expensive switching equipment.

As part of the public switched network, centrex is a constantly growing and developing system. New features are being added on a continual basis. These services include selective call waiting, selective call screening, and access to circuit and packet switching services, which will be explained later. City-wide centrex is another future service.

Alarm Service

Alarm service is another type of service available through the public network. It uses the subscriber's existing telephone service to transmit alarms (such as fire, theft, and medical alert). Monitors are set up to scan the premises. Lines connected from the monitors transmit signals to the alarm company.

These types of services can be provided on a dedicated line for continuous monitoring or by sharing the local phone line. When using the dedicated line, constant monitoring of the customer location by the alarm company is possible. Under this arrangement the alarm company is usually able to identify if the line is cut. However, dedicated service can be expensive because cost is distance sensitive. As the distance between the customer and the alarm company increases, so does the cost.

The option of using the existing customer phone line is much more economical, but there are trade-offs. The alarm location is not provided with constant monitoring. If the line is cut, there is no alarm sent to the monitoring agency, and protection service is lost.

Upgrades in the metropolitan area networks are solving this problem and reducing costs. A new technology called derived channel gives both the economy of shared use and the benefits of dedicated lines. Derived channel will allow simultaneous voice or data on a line being used to constantly monitor the premise. With the two signals piggybacked on the same line, the customer's location will be under constant surveillance, and the alarm company can be immediately aware of any interruption in the signal.

The service is now limited to alarming, but its growth potential lies in expanded use for such things as continuous premises monitoring of power consumption and control of environmental factors (heat, humidity). These now require a separate computer and interface for each service.

Data Communication

There are also many types of digital services for enhancing computer data communications. When transporting voice over the telephone lines, the conversation is almost continous with only short pauses. When interactive computer data is transmitted, it occurs in small, high-density bursts of information. A majority of time the circuit is idle. A Public Packet Switching Network (PPSN) uses a more efficient method of transmitting computer data. PPSN is a communication network that uses packet switching technology, digital transmission facilities, and existing loop plant. It provides economical common-user transport for bursts of data within a local serving area. In addition it provides access to interexchange carrier's packet networks for transmission of data outside the local network.

Packet switching. Packet switching uses an entirely different switching technology from the type used in the analog voice network. It more efficiently uses the transmission and switching facilities for data transfer. There are no permanent circuits established. The transmission circuits are only intact for the time it takes to transmit the data. In packet switching, information is stored in discrete packets using a format or protocol called X.25. This X.25 packet of information is a section of serial bit stream that is buffered and divided. It can vary in size and contains all the data needed to route the packet through the network, such as place of origin, destination, sequence in message, and error checking.

The first step in sending data over a packet switching network is to establish a path for transmitting the information. A message is then sent out, and the destination computer sends back an acknowledgement indicating completion of the circuit. Thus a permanent virtual circuit between the two computers is established. The initiating computer next enters the transfer mode. Data is divided into packets and sent. At the first switch the packet is examined. Once its destination is determined, it is sent to the next switch. This process is continued until the destination computer acknowledges receipt of the packet. The only time the network facilities are being used is when the packets are actually in transit between sender and receiver.

Packet switching has distinct advantages over using conventional phone lines. The establishment of a virtual circuit causes no facility usage or reservation. By using packet switching for "bursty" data, transmission resources (e.g. trunks) are more efficiently used, since they are dynamically allocated only when actual data is sent and are free the remainder of the time. The entire process is transparent to the user, who has the perception of exclusive use of the resources. Digital transmission is provided throughout, and the error-checking nature of X.25 protocol almost eliminates undetected errors.

Protocol conversion. When transmitting data over a network, two types of data transmission can be used:

> *asynchronous*—stop-start random transmission with each character having its own limited error-checking "bits."
>
> *synchronous*—predictable and timed transmission of a number of characters at the same time, with *extensive* error-checking capabilities.

A packet switching network uses the synchronous form of transmission (X.25). Since many data terminals use asynchronous transmission, a conversion from asynchronous to synchronous is necessary. This conversion to the X.25 synchronous form is called protocol conversion.

The FCC has given approval for the former Bell System Operating Companies to offer protocol conversion from asynchronous to synchronous. When the public network provides this conversion, users can enjoy the economical benefits of packet switching without having to purchase protocol-converting equipment at each operating station. The conversion can be performed at a central location in the network. This service positions the public network economically to meet the sophisticated data transmission needs of customers who are connected to this network.

Public Switched Digital Service. Another digital service under development for use in digital metropolitan networks is Public Switched Digital Service (PSDS). PSDS provides end-to-end digital connection through the network at rates up to 56/64 Kbps. PSDS features high-quality, cost-effective transfer of moderately heavy data among multiple locations. The high bit rates make bulk data transfer, high-speed facsimile, computer graphics, and encrypted secure voice possible. In addition the dialup line is still available for normal voice communications when not transmitting data.

PSDS, unlike packet switching, will not require an entirely different communication technology. It will be able to use the already existing network of digital switches, analog switches with digital connectivity, and digital trunk facilities. The customer will need digital access lines, which will extend directly to his premise. Customer premise equipment will also need rate-adapting electronics to synchronize it to the 56/64 Kbps access line if the customer wants to transmit slower speed data.

High-Capacity Digital Service. The digital service that now offers the highest available data rate is High Capacity Digital Service (HCDS). This type of end-to-end digital service is on the DS1 or 1.5-million-bps level with the potential for even higher rates in the future. Although the service provides enormous capacity, it does have some limitations. HCDS is a hard-wired circuit. This means it is installed between predesignated locations and does not have any switching capabilities. In the future, as digital switching technology develops, HCDS will eventually become a switchable signal circuit and serve as one of the basic ISDN elements.

DIVESTITURE

The most publicized event in telecommunication history was the divestiture of AT&T from its Bell Operating Companies (BOC) on January 1, 1984. The world's largest corporation was separated into AT&T and seven regional holding companies. The holding companies contained the 22 former Local Operating Companies (LOCs). The breakup's most profound impact was on the division of service responsibility among the now-separate entities of the formerly integrated Bell System.

BOC Responsibilities

The BOCs will be responsible for the traffic in a specific geographical area called a Local Access Transport Area (LATA). LATAs were developed as a result of divestiture. Within the boundaries are located areas of common economic interests. The BOCs will provide dial tone, local calling, and toll calling within the LATA. This type of traffic is called intra-LATA. Any traffic outside the LATA is called inter-LATA and will be handled by long-distance carriers, such as AT&T, GTE, MCI, and SBS.

Equal Access to Long-Distance Services

Prior to divestiture customers could choose different carriers for handling their long-distance calls, but it was easier to use AT&T because of the dialing procedures in effect at the time. AT&T could be selected by merely dialing a 1, the area code, and then the number. To select another carrier required a much more involved process of calling the carrier, dialing an access code, and then dialing the desired party. Following divestiture, all this changed as the new concept of equal access was born.

Simply stated, equal access gives the customer equal access to and a choice of long-distance carriers. When the serving central office converts to equal access, notification of the conversion is sent out to all the customers. At this time they must make a choice of which long-distance carrier they want to provide service. Then any time a customer dials 1 to make a call, inter-LATA, it will be automatically handled by the selected primary carrier. Anytime the customer wants to use a different carrier to handle a particular call, the customer can dial a five-digit code identifying the chosen carrier, and that carrier will handle the call. The primary carrier will remain unchanged and will handle any subsequent call unless another carrier is again selected by dialing a five-digit identification code. A customer who wants permanently to change primary carrier can do so by notifying the local telephone company.

Equal access is only one of the many advantages created for the customer by divestiture. There is now a much wider range of communication vendors and equipment available. More service providers are appearing in the market, giving increased options and enhanced features of telephone, data, and video services.

TELEPORTS AND TELEHUBS

A teleport deals with the connecting of satellite communication facilities with intelligent buildings in a real estate development. A broader application, or a telehub, applies the teleport concept to a more general geographic area that has access to sophisticated communication services through a distribution network. This distribution network must be capable of meeting all of the user's communication needs between the office site and the connection point to the high-capacity facilities. Typically the hub provides access to microwave, satellite, and fiber optic facilities that serve as the vehicle to transport voice and data over long distances.

Some of the most visible applications of the teleport concept are large commercial real estate projects. Examples of these are the Staten Island project in New York and the Bay Area Teleport in San Francisco. The idea behind each of these projects is to enhance the real estate development through shared use of communication facilities, which might prove too costly for individual users. By sharing the use and cost, they become economically attractive.

The metropolitan area networks will continue to play a key role by providing access through telehub type of arrangements to sophisticated, high-capacity telecommunication facilities. Under this telehub concept, the public network will provide access to the intercity facilities over high-capacity lines. Telecommunication services can be provided via fiber optic, coax, microwave, and many other types of transmission facilities. By accessing the high-capacity systems through the telehub, the end user does not necessarily need dedicated lines to teleport-like facilities. Costs may only reflect actual use of the facilities. Other benefits of the switched network, such as diverse routing, constant monitoring, and 24-hour maintenance, would also be available. The important factor is to provide the user with a network that can meet specific informational and service needs at the most economical price.

Philadelphia—The Telehub

The Philadelphia communication network is ideal for providing telehub services. It has all the digital features of a typical metropolitan area network and is a key factor in serving and attracting businesses to the Philadelphia area. It is a dynamic network constantly growing and evolving, and typifies the standard of excellence found throughout the public switched system. The success of this network can be found in its ability to meet the local business community needs, both present and future, and translate them into working communication systems. It is the resource that links the city and business tenants with the Information Age.

The Philadelphia communication network can provide the latest digital services: 95 percent of subscriber plant has the capability for data speeds over 144 kilobits per second; 100 percent of interoffice routes are equipped for 1.5 Mbps transmission. In addition 100 percent of the switching is Stored Program Control (SPC), and 1984 saw the completion of Philadelphia's first lightwave ring capable of providing high-capacity optical digital service directly to the customer. An additional ring has been installed, and more are scheduled for installation in the future. With all these digital facilities, Philadelphia is rapidly approaching an end-to-end public digital network, or ISDN.

One of the first essential elements of an ISDN is digital transmission facilities between the customer and the central office. In Philadelphia, as in many other cities, the plant is well suited for providing this connection. As previously mentioned, 95 percent of the customer serving links are capable of providing 144 Kbps service. This data rate, which has the capacity to transmit more than five typewritten pages of text per second, can also handle the ISDN 2B + D channel configuration. The close proximity of customers to the central office is one reason for such a high percentage of availability.

Another vital element to building an ISDN is digital interoffice transmission facilities. The entire downtown area is blanketed with digital carrier routes. The basic transmission rate of these facilities is 1.5 Mbps, enough for a large company to place a headquarters computer in Philadelphia and communicate with branch offices. The capacity of these systems will expand well above the 1.5 megabit rate with the increased use of fiber optic lightwave systems.

The Philadelphia metropolitan area is also continually upgrading its switching capability. It has already converted all of its electromechanical switches to modern electronic switches. These electronic switches are computer controlled and can be programmed to provide new types of services. One of the essential building blocks of an all digital system is electronic digital switching. As development toward an ISDN continues, more and more digital switching facilities will come into use. Because of the close proximity of customers to central offices and the increasing availability of digital switching, the center city area will be ready to give business customers the new services they want.

The importance of these new digital capabilities is the service they will provide to the business customer and high technology tenants. Whether the customers are the government, banks, hospitals, universities, insurance companies, law firms, manufacturers and distributors of goods and services, retailers, wholesalers, or any other type of business, the metropolitan area network can meet their growing informational needs. For voice communications, the digital network can provide custom calling, call waiting, call forwarding, speed calling, and three-way calling. With the greater reliance on computer systems, the digital network is ideal for data base inquiry/response, videotex, teleconferencing, electronic message systems, online poling, time-share computer terminals, energy and environmental controls, high-speed computer data transfer, and remote utility meter reading.

All these services further enhance the Philadelphia downtown area and serve as a valuable resource in attracting new tenants and increasing real estate values. Philadelphia took another step toward becoming a major east coast telehub in 1983 and 1984 by connecting with several northeast corridor lightwave systems. These systems serve major east coast cities and other long-distance carriers.

As new technologies and telecommunication services become available, the metropolitan area networks will be the first to capitalize on these new capabilities. With the telehub arrangement, the entire spectrum of telecommunication services will become available through the developing metropolitan area digital networks as they move closer to the ultimate goal of an ISDN.

Authors

G. William Ruhl
Division Manager
Bell of Pennsylvania
Philadelphia, Pennsylvania

William Ruhl is the division manager for Network Services and Outside Facilities Engineering for the Central Area of Bell of Pennsylvania. In that capacity, he is responsible for the engineering of the cable and fiber optic facilities and the maintenance of the central office switching systems and associated transmission facilities in central Pennsylvania. Prior to this assignment, he has held positions in which he was responsible for the planning of operations systems, technical support, and central office and interoffice modernization for the company. Ruhl is a graduate of Lehigh University with a degree in electrical engineering. He participated in the Bell Laboratories with a degree in electrical engineering. He participated in the Bell Laboratories Operating Engineers Training Program and was an instructor in long-range technical planning at the Bell System Center for Technical Education.

Randall C. Frantz
Assistant Planner—NSI
Bell Atlantic
Philadelphia, Pennsylvania

Randall Frantz joined Operator Services Planning Staff for Bell Atlantic's Network Services Incorporated in 1985. As an associate planner, he develops long-range plans for Regional Operator Services. Prior to transferring to Bell Atlantic, he worked for Bell of Pennsylvania where he entered the management training program in 1983. While with Bell of Pennsylvania he worked in several areas; Outside Facility Engineering, Carrier Access Billing, and Network Design. Frantz graduated from the United States Air Force Academy in 1977 as a Distinguished Graduate with a BSEE.

Chapter 18

Teleports: The Real Estate and Planning Connection

Piero Patri
Whisler-Patri

Michael J. Harlock
Whisler-Patri

Alan D. Sugarman
Merrill Lynch

Peter Valentine
Comsul, Ltd.

Outline

Although to most people the word *teleport* brings to mind Ray Bradbury novels, domed Venusian landscapes, or telepathic twins coping with the communication dilemma of light speed, to the communications industry a teleport is no more than an open field with clusters of satellite dishes. These antenna farms, as they are often called, have been with us since the first communication satellites were lobbed into orbit.

In the narrowest sense, a teleport is any facility for satellite-bound transmissions, or uplinking, and may consist of no more than a single dish on a high-rise rooftop. Satellite dishes, however, are usually just one highly visible component of these antenna farms, especially as they have evolved into the teleports of today.

Teleports in their emerging form combine antenna farms, fiber optic and microwave networks, and so-called smart office parks and/or smart buildings. The combination of these communication facilities with real estate is receiving increased attention from real estate developers and investors. The development of teleports has become a way to create real estate value.

The interest of the real estate industry in teleports is due in part to the increasing importance of communications to the service and information economy which megathinkers say is upon us. With the explosion of information technology in the 1980s, the appetite of corporations, institutions, and government for the transmission and digestion of greater and greater amounts of voice, data, and video over longer distances has established the fundamental importance of communications to modern business. Access to communications is now perceived to be an important component in the value of real estate. Business need for efficient and economic telecommunications has begun to influence real estate decisions.

Developing a teleport demands not only the technical sophistication to design, construct, and maintain complex communication facilities, but also the application of sound real estate and development planning and an understanding of the market for telecommunications-enhanced real estate. Fortunately, we can make an analogy to ports and trade routes in the earlier phases of industrialization, and we can learn planning and marketing lessons from more traditional real estate developments. First, however, we will discuss the issue of the centralization versus decentralization of business organizations in the information age.

CENTRALIZATION OR DECENTRALIZATION

Anthony Downs of the Brookings Institution, in an article entitled "Future Impacts of Telecommunications on Real Estate and Urban Growth," (presented to the Landtronics Conference organized by George Lefcoe), has postulated that new age technology, like the telephone before it, is not so much a determinant of the location of economic activities as it is a permissive agent. Simply, new technology allows wider choices in where we live and work. He believes that the more traditional factors dictating location, such as governmental regulation, housing supply, and transportation systems, will continue to be the primary ingredients in business location decisions. Telecommunications give organizations choices in responding to these factors as they expand their need to communicate. In this sense, the trend has and will continue to foster decentralized institutions. The front-office/back-office split is one clear example stemming from the increase in central business district rents in conjunction with the availability of moderately priced employee housing in outlying areas. Another manifestation of decentralization is the ability of individual workers to telecommute from their homes or neighborhood work centers, assisted by distributed computing and advanced telephones.

While teleports that allow companies to take advantage of many domestic or international branches can be seen to be an extension of these decentralizing forces, they are simultaneously a force for bringing businesses together to overcome the capital and operational costs of riding the digital highways. GTE's Office of Enhanced Real Estate in Los Angeles has stated that its current strategy is to encourage information-hungry businesses to congregate into smart parks along their "Olympic" fiber trunk, with connection to one or more teleports expected in the near future. Ten such parks are expected to open by the end of 1986.[1] In West Los Angeles, Whisler-Patri is exploring the feasibility of a teleport development for the City of Oxnard. Fiber optic pathways may have the same growth-inducing impact as the interstate highway system did in the 1950s and 60s. Railway rights-of-way, now being pursued for long-distance fiber lines, may once again act as a magnet to real estate development, as may previously overlooked features such as the California Aqueduct being studied by the California Department of Water Resources for a fiber link between Northern and Southern California.

Teleports planned as enhancements to real estate derive from a more comprehensive appraisal of the nature of the communication business. As the telephone company realized several years ago, a successful communications medium must span from first mile to last to form a conduit that is simple and reliable. The teleport proposes to reach out and touch users at both ends offering service, flexibility, and dependability. For the near future at least, whether an information-based company keeps its urban roots or starts packing will depend in part on the presence of an integrated urban network linking users with each other and the satellite gateway.

THE PORT OF THE INFORMATION AGE

As seaports, railroad stations, and airports brought the common carriers of the Industrial Age together for the convenience of shippers and travelers, teleports now provide a similar function in the Information Age. Ports have always conferred urban status, reflecting a mass of users whose business required interaction and exchange with other metropolitan centers and their associated exurban areas. Teleports are one means of achieving critical mass information and telecommunication services. Although conceived initially as the focus of satellite communication, their function is far more extensive today, including terrestrial networks such as fiber optic networks. However, at present, satellite transmission facilities may be the most visible component of the teleport. The importance of the teleport may be in its concentration of comprehensive telecommunications services, some of which cannot be offered in the central city. For a number of economic, technical, political, and environmental reasons, placement of a massive satellite farm in the middle of a modern city is as feasible as placing an airport for Concordes in Central Park. To the extent that satellites will remain a key component of 21st century communication, the need for protected transmission facilities will continue to influence the location of real estate developments offering comprehensive telecommunication services.

Many may dispute this thesis. Some contend that fiber optic technology will make satellite communication obsolete or that fiber optic networks will permit users to *not* locate at the teleport, making the real estate development at the teleport unnecessary. However, even the most pessimistic critics of satellite communications and teleports acknowledge that satellites will remain a key communication method for the next 10 to 15 years, and even longer for point to multipoint video communication. Moreover, there are economic and technical advantages for many activities to locate at the information node. It is the premise of the teleport as real estate that established communication-oriented business centers will maintain their critical mass, even after the demise of the original raison d'etre.

In considering the impact of technology on land uses, it is useful to examine some of the lessons learned from the evolution of airports. The first airports were created spontaneously to support a technology that rapidly proved its worth in servicing existing economic enterprises and habitational patterns. Often, their location was a short-sighted compromise between central-city proximity and available open land. As we have seen, many airport sites have proven difficult to adapt to the mushrooming growth in air transportation (and the growth of cities and suburbs) without having a negative impact on neighboring uses.

Where airports have had the luxury of surplus acreage, an elaborate collection of uses has evolved. For business travelers, an extensive support system has evolved that eliminates the need to undertake the vexsome process of traveling to the city center. Executive suites, sushi bars, and convention complexes are all within a shuttlebus ride of the departure gate. With this agglomeration of facilities serving the businessperson, it was only natural that new and expanding businesses whose livelihood depended on frequent access to air carriers would begin to locate around these centers such as spare parts distributors to facilitate the delivery and receipt of air cargo. In many cases, airports have become cities unto themselves, with their own concentrations of office towers, religious disciples, and air pollutants.

Airports thus evolved from being places from which one embarked on an air voyage to engines generating their own economic activity centered around the airport. Unfortunately, few airports were planned with this in mind. Ground transportation within and to airports was inadequate, parking was frequently an afterthought, and the location of support industries was not even a consideration.

We predict that teleports may serve similarly as concentrators of communication-related businesses notwithstanding the lukewarm response of the market to this new concept. The future of the teleport requires careful planning to meet both its public goal of economic development and its private goal of real estate development for economic profit. Thus, we must consider the physical planning criteria equally with the evaluation of potential real estate use.

TELEPORT PLANNING

Several criteria are vital in picking locations for a new teleport-cum-business park. The site must, above all, be practically free of radio frequency interference and microwave transmission paths. Big trees and blocking hills could be a detriment. The land must allow visibility of all North American communications satellites and as many international satellites as possible. In most areas of the United States this will not be an obstacle since satellites have been parked where they can transmit to most of the country. Environmental screening for aesthetics as well as for the real and perceived dangers of microwaves is a must. Instantaneous access to both short- and long-term backup power supplies must be assured. Connection to multiple regional power supplies may be important. The teleport site should facilitate superior security surveillance with one controlled access point and 24-hour perimeter patrol.

Master planning for the teleport business park should tackle the same site constraints and opportunities that would be considered in any quality development of 100 or more acres, probably the least amount of land required for this type of venture. Landscaping, building setback, parking, and architectural design parameters must be codified to maintain the quality of future development. The park's utility trenches, in addition to water, sewer, and electrical lines, must provide sufficient conduit capacity to support a high level of network cabling between buildings and the antenna compound. Whisler-Patri's experience in planning the Lincoln Technology Center in San Jose, California, demonstrates that this capacity for flexibility can be installed at a relatively low cost.

WHO WILL TELEPORTS SERVE?

Although technical requirements may be difficult to meet, the first step in teleport planning must be a careful analysis of the potential user base and a realistic appraisal of competing technologies and facilities. The form this analysis should take brings into focus the difference between public/ private ventures and solely private developments. However, existence of a viable private market is essential to both types of ventures. The key to any real estate venture successfully enhanced by communications is a product tailored to the needs of targeted customers.

Teleport Benefits

The teleport with its associated terrestrial networks offers a number of advantages to business communication users.

Flexible access to sophisticated communication. Multiple dish antennas and terrestrial networks can be accessed easily, depending on the user's needs and budget.

Low capital investment for satellite communication. Satellite signal scheduling may be continuous, intermittant, or for special events. For users in the two latter categories, capital investment in dedicated facilities would probably be uneconomic. The same may be true for continuous users, for the capacity of a transmitter permits numerous continuous users to use the transmitter simultaneously.

Access to alternative telecommunications networks. Initial users of a teleport may be attracted to its terrestrial network, permitting sidestepping not only the local Bell operating companies but also the tangled microwave networks of other bypass carriers and the microwave congestion of inner cities. In such cases, the teleport also serves as a communication switching node, moving traffic through its own network as well as facilitating interconnection with other networks.

Backup for traditional pathways. Commercial common carriers are also expected to continue using satellites, even after long-haul fiber optics highways are completed, to ensure dependability at least the equal of Mother Bell.

Maintenance. A teleport offers an owner of satellite dishes insurance and maintenance of the equipment 24 hours a day by teleport management.

Consolidation of telecommunications staffs. Some users may value the ability to consolidate their technical staff at one place, instead of maintaining several diverse locations.

Secure electric power. A well-planned teleport location also provides sophisticated alternative power sources to protect against short- or long-term interruption. For example, the Staten Island Teleport overlaps both major power grids of the northeastern United States; this is as much redundancy as any location could offer. As an added benefit, in the case of the Staten Island Teleport, relocation from Manhattan will also result in a 35 percent reduction in a company's total electric bill.[2]

Access to satellite, microwave, and fiber optic service organizations. Once a teleport is in operation, vendors and maintainers of hardware may choose to locate at the teleport site because of the location of sophisticated equipment requiring no down time. These businesses will attract their own customers.

Ability to transmit and receive large amounts of information to and from customers. Certain businesses that depend on the transmission and receipt of voluminous video or data may find that on-site location with flexible communication facilities is advantageous and permits better control.

Secure physical environment. A properly planned teleport can provide essential security for data processing and storage organizations requiring absolute confidentiality.

Lower cost labor markets. If properly located, an exurban teleport may offer access to lower-cost or higher-quality labor markets. This may be attractive to telemarketing organizations or back-office organizations.

Potential Users of Teleport Real Estate

What kind of company would seek a teleport location? Following is a sample of the types of operations that may find a teleport location attractive.

Data processing back offices. The operations of financial services organizations would benefit from most of the advantages listed above.

Video broadcast distribution centers. There are a number of private and not-for-profit video networks that may well consider a teleport location for their distribution centers. Video broadcasts could be controlled and distributed anywhere in the world. There may be fiber optic links to production centers located in urban areas or even across the country.

Teleconferencing. Video teleconferencing networks may require sophisticated and flexible control centers. Companies managing and providing these services are logical teleport candidates.

Backup data storage facilities. Most major companies have a need for multiple and continuously updated locations for storage of computer data in case of a computer failure or natural catastrophe. Private companies frequently offer this service. Data could be transmitted to the secure teleport environment by satellite, fiber optic, or tape. In the event of catastrophe, data could be efficiently transmitted to the selected backup processing center.

Corporate communication centers. Most large corporations have established or are establishing their own private communication systems encompassing voice, data, and video. Teleport real estate represents an excellent choice for the location of the primary control centers for these networks.

Message and telemarketing centers. These organizations require a combination of low-cost labor and flexible and cost effective communication. Teleports, located near large suburban labor pools, are an excellent choice for these organizations.

Telecommunication hardware service organizations. Accessibility to users is obvious.

Information service providers. At present, information database services are primarily providing text information over voice grade telephone. These services require large computer centers and extensive communication capacity. In the future, it is expected that information databases with associated facsimile databases will look to satellites to meet the demand for wider bandwidth transmission.

The foregoing are potential initial users of teleports. Once established, other tertiary businesses will be attracted to teleport real estate, including restaurants, retail stores, and hotels as well as traditional office organizations.

TELEPORTS AS EXTENSIONS OF SMART BUILDINGS AND PARKS

Teleports can provide an extension for the smart building, expanding the bypass technology inherent in the shared-tenant concept to long-haul data, voice, and video communications. Some corporations have been utilizing their rooftops for dish installations, but problems inhibit widespread application; not the least is cost. Single uplinks can cost in the range of $250,000 to $500,000 (not including satellite transponder rental), and information-intensive consumers often require multiple satellite access to guarantee transmission continuity. Besides the capital expense of such equipment, sufficient rooftop space is already becoming scarce in the largest cities, especially on the new tapered towers. Building owners are also charging monthly rents of as high as $10,000 for rooftop antenna space. Perhaps the least controllable aspect is the level of radio and microwave interference in many central business districts. Finally, design review boards are increasingly sensitive to high-tech gadgetry that is not fully integrated with office building design.

For these reasons, the logical extension of smart buildings (or wide-rise smart parks) is a land link to a central, protected satellite transmission site.

From this perspective, the teleport's terrestrial network appears to be competing with its real estate development, providing satellite accessibility to remote users who might otherwise move to the teleport business park. However, one inducement for on-site location is reduced or eliminated terrestrial charges. For a 100,000-square-foot tenant with demand for 1,000 T1 lines, at Harbor Bay Teleport, savings in monthly carrier charges can be as much as $23,000 (based on current tariffs of the California Public Utilities Commission).[3] Moreover, as indicated previously, many companies will seek on-site locations for other compelling reasons.

TWO TELEPORTS

The first teleport, and the best known, is the Staten Island Teleport that popularized the name *teleport*. This ambitious development, with substantial financial and political backing, includes a comprehensive real estate development, a state-of-the-art satellite facility, and an extensive fiber optic network extending into the boroughs of New York City and into New Jersey as far south as Princeton. It has developed roads and utilities to serve over 200 acres of office building sites, all with direct linkage to an array of satellite dishes that may be owned individually, collectively, or maintained for common use by the Teleport Communications Co., a joint venture of Merrill Lynch and Co. and Western Union.

As a public/private venture jointly sponsored by the Port Authority of New York and New Jersey and the City of New York, the goals of the Teleport were considerably grander than the enhancement of 200 acres of suburban real estate. With recent stories in mind of corporations deserting the New York metropolitan area, Teleport planners saw the project as a vital inducement to business retention and growth. Key to this strategy was the construction of a 52-mile fiber optic network linking the Teleport with downtown Manhattan as well as subcenters in Brooklyn, Queens, and New Jersey. Fiber optics allowed the Teleport to be constructed in the flatlands independent of line-of-site microwave relays, the typical terrestrial link to Manhattan office buildings. Absence of microwave interference allows unencumbered C-band transmission, the more dependable of two frequency bands authorized by the FCC for satellite communications. (The smaller and more mobile Ku-band dish equipment currently is subject to signal loss in heavy rainstorms, a crippling disadvantage to service-sensitive users.) Even in the relative radio tranquility of Staten Island, a 50-foot earth- and metal-clad wall was erected to screen the 11-acre dish compound to ensure failsafe operations now and in the future.

Another teleport, located on the opposite side of the country, is Harbor Bay Isle, a large mixed-use office and R&D development directly across from the bay from San Francisco. This development has won zoning approval for a 347-acre teleport site nine miles away in an undeveloped valley, providing full view of all North American communication satellites. Microwave relays link the Harbor Bay Teleport, its office park, and 20 other locations throughout the Bay Area to points as far as Sacramento (approximately 90 miles). Harbor Bay Isle has advertised that its teleport facility can serve the entire Northern California region.[4] However, other dish installations along the microwave pathway will continue to compete for this business. The point is that for smart parks like Harbor Bay, a co-location at the antenna compound is neither feasible nor highly relevant.

These two teleports are examples of a public/private venture on one hand and an exclusively private venture on the other.

PUBLIC, PRIVATE, OR BOTH

The Staten Island Teleport is an instructive example of public and private partnership in teleport planning. With the Port Authority of New York and New Jersey as major sponsor and developer and the active involvement of the City of New York, the objectives for New York's teleport were clearly regionwide. While a large amount of business is expected from the 1 million square feet of tenants housed in the development, the vast majority of companies are now predicted to use the Teleport by tapping into the fiber optic network from remote facilities (particularly in Manhattan). Advantages in public sponsorship include the elimination of political barriers to the Teleport and terrestrial network as well as an implied public blessing to bypassing the local telephone company, thereby avoiding potential regulatory snags. Problems have centered on the characteristic slowness of governments to do anything. In New York, the complexity of contractual arrangements have significantly delayed the start of service. An agency's five-year strategic plan can quickly lose relevance in a technological environment characterized by one-year product runs.

Private ventures, such as Harbor Bay Isle, can be agile in reacting to the shape of the marketplace. However, they will almost certainly be more limited in objective, given the tremendous capital outlays in support of what is largely an unproven demand. For this reason, Harbor Bay Teleport will not purchase its own dish equipment but will develop, lease, and maintain sites for equipment owners.[5] Time saved in organizing the private team may be offset by political and zoning obstacles, especially in states with tough environmental legislation. Even when health concern about microwaves is successfully countered, the 30-foot diameter dishes or their screening wall may be viewed as a technological monstrosity to nearby residents.

HOW MANY TELEPORTS PER REGION?

The big unknown for both public and private developers is the latent market strength and appropriate timing for the teleport concept. For this reason, answers to such questions as who will utilize the services and how many teleports one city can support tend to be vague and laced with marketing hyperbole. At least one urban center, Houston, has three teleports/business parks in active development. Jay Watson of Watson Communications, owners and managers of dish antenna sites around San Francisco, believes Bay Area geography will dictate at least three major teleports. Certainly, separating fact from promotional fantasy in the teleport business will be difficult for some time to come.

Still, many people entrusted with boosting local economies feel that satellite access will be an increasingly important carrot in luring big employers and keeping old friends. The Downtown Central City Association of Los Angeles is pursuing teleport developments in the Los Angeles region partly out of civic concern for the volume of Pacific Rim communications that leaves California via San Francisco's submarine cable, but primarily to maintain the city's successful track record in attracting growth industries and jobs.[6]

■   CONCLUSION   ■

Whether at the teleport or integrally linked to it, the value of real estate accessible to the satellite gateway is unquestionably enhanced. The important question for some time will be, by how much? Careful planning for the marketplace now and in the foreseeable future must accompany traditional, physical planning analysis. The first ports have opened for information business. The most interesting information to the real estate community, however, will be the reaction of the first customers.

■   NOTES   ■

1. Bob Lavallee, Director of Enhanced Real Estate, GTE of California, Newbury Park, CA.

2. Richard Kennedy, Cushman & Wakefield Inc., New York, NY.

3. John Harrison, Harbor Bay Telecommunications, Alameda, CA.

4. Ibid.

5. Ibid.

6. Ibid.

Authors

Piero Patri, FAIA
President
Whisler-Patri
San Francisco, California

Piero Patri, FAIA, is president of Whisler-Patri, a 100-person architecture, interior design, and planning firm located in San Francisco. Since its founding 31 years ago, the firm has been responsible for the design and planning of over 20 million square feet of office space. Clients include developers such as Gerald Hines, Marathon U.S. Realties, and Lincoln Property Company and major corporations such as Xerox, Charles Schwab & Co., and Chevron. Whisler-Patri has over 2 million square feet of "smart" office space in design or operation and has overseen the design of a 450-room "smart" hotel.

Patri is in his second year as chairman of the National Research Council Committee on Technologically Advanced Buildings. The Council is the research arm of the National Academy of Science and National Academy of Engineering, and the Committee has been charged to evaluate the application of office automation and enhanced communications in government facilities as well as in the private sector.

Patri is a member of the Urban Land Institute's High Tech Committee. He served on the Publications Review Committee for the ULI's *Smart Buildings and Technology-Enhanced Real Estate* and is a contributing author to their forthcoming publication, *The Changing Office Workplace.* In addition, he recently coauthored a chapter on "Retrofitting Existing Facilities to Accommodate Office Automation" for *High Tech Real Estate,* published by Dow Jones-Irwin.

Patri frequently speaks before business, government, and professional groups on the topic of "smart" buildings and office productivity. Some of these groups include the Industrial Development and Research Council, the Urban Land Institute, the Congressional Office of Technology

Assessment, and recent business conferences in London, Philadelphia, New Orleans, Washington, and San Francisco.

Patri was educated in Milan and Florence and graduated Phi Beta Kappa from the University of California at Berkeley. A native San Franciscan, he served as a Field Instructor in Carrier and Repeater Communications for the U.S. Army Signal Corps.

Michael Harlock, AIA, AICP
Associate
Whisler-Patri
San Francisco, California

Michael Harlock has 14 years experience as an architect and a planner. He recently oversaw the site selection and approvals processing relative to establishing the GTE Mobilnet system in the San Francisco Bay Area. His current involvements include assisting the City of Oxnard in developing and implementing a city-wide telecommunications system, master planning an 800,000 square foot R&D center for Xerox Systems Group in Sunnyvale, and a 60-acre hotel, office, and commercial development, also in the Silicon Valley. Harlock was educated at the University of Michigan, Ann Arbor, and at the University of California, Berkeley.

Alan D. Sugarman
Vice President
Merrill Lynch
New York, New York

Alan D. Sugarman is vice president and associate general counsel for the real estate and real estate financial services group of Merrill Lynch & Co. Prior to assuming that position, Sugarman served as the general counsel of Merrill Lynch, Hubbard Inc., the group's institutional real estate division. While previously engaged in private practice in New York City, he specialized in litigation, corporate, and real estate law. Sugarman was also general counsel, Roosevelt Island Development Corporation, and senior staff counsel, INA Corporation. He is engaged presently in the practice of real estate, finance, investment, and securities law. In addition, he has considerabl. experience in computer and telecommunications law and applications and other areas relating to technology. He is a graduate of the University of Chicago Law School, where he was a member of the *University of Chicago Law Review.* He also holds a bachelor's degree in electrical engineering from Tufts University and is a member of the Eta Kappa Nu and Tau Beta Pi engineering honor societies. He is the coeditor of *High Tech Real Estate,* Dow Jones-Irwin, 1985.

Peter B. Valentine
President
Comsul, Ltd.
San Francisco, California

Peter B. Valentine is president of Comsul, Ltd., a telecommunications planning and consulting group founded in 1967. The organization has successfully completed over 1,200 communications design projects over the past 17 years. Valentine came to Comsul in 1970 from the investment and securities industry. His background in business management brings a balance of economic common sense to the firm's technical expertise.

Chapter 19

Telecommunications Systems and Large World Cities: A Case Study of New York

Mitchell L. Moss

Graduate School of Public Administration, New York University

Outline

INTRODUCTION

The rapid development of teleports is largely a product of two phenomena: the emergence of selected metropolitan regions as "information capitals" and the opportunities to enhance a region's telecommunications infrastructure facilitated by deregulation and technological advances. While a teleport represents a new and potentially significant element in a region's telecommunications infrastructure, it is also important to recognize how a teleport fits into the overall pattern of telecommunications development within a large city. This chapter provides a case study of the telecommunications infrastructure in the City of New York and explores the implications of telecommunications for urban economic development. The chapter consists of four parts: (1) an analysis of the city's information sector, (2) an assessment of the city's telecommunications infrastructure, (3) a description of the telecommunications systems within New York City, and (4) policy options for using telecommunications in economic development.

NEW YORK CITY'S INFORMATION SECTOR

The current strength and growth potential of New York City's economy depends substantially on the city's function as a headquarters site for major corporations, as a center for publishing, television, and other media, and finally as a capital for such information-based services as financial services, banking, law, management consulting, accounting, and advertising. These services have gained prominence during the past 25 years as the city's economy shifted from one of goods production and handling (i.e., manufacturing, trade, and transportation) toward one characterized by a concentration of information-handling activities. In 1958 the information sector—18 of the 51 private nonagricultural industries—accounted for 35 percent of the city's private employment. By 1984 the information sector had grown to 55 percent of the city's private employment, as illustrated in Exhibit I.

Furthermore, the information sector accounted for 41 percent of the city's income in 1958 and 58 percent of the city's income in 1984. These figures do not include federal, state, and local government agencies, which also have a high information component. Banking, securities, and business services together accounted for 23 percent of the total city value added in 1984, more than double that in 1958. These industries are highly information intensive.

Information handling and processing is not confined to information-based firms, but impinges upon all economic activities. This is particularly true for New York City with its strong headquarters concentration. For example, many firms engaged in goods production (e.g., apparel manufacturing) maintain their top management and support staff in New York City. Occupational

EXHIBIT I
Percentage of New York City's Employment and Income in Information Intensive Sectors

	Employment*		Income†	
Industry or Group	1958	1984	1958	1984
Printing and publishing	4.0	3.2	4.2	3.5
Instruments and electrical machinery manufacture	2.4	1.1	2.7	1.2
Communications	2.5	2.9	3.2	5.0
F.I.R.E.‡	11.9	17.3	17.1	23.6
Selected services§	14.1	30.6	13.6	24.4
Total Information Sector	34.9	55.1	40.8	57.8

* Percent of total private non-agricultural employment.
† Percent of total value added of private non-agricultural establishments in constant dollars.
‡ Consists of banking, credit agencies, securities, insurance and real estate.
§ Consists of business services; motion pictures; amusement services; health, legal and educational services; social services; nonprofit organizations; miscellaneous services and museums.
Source: Drennan, 1985.

employment statistics in fact show that in five of the six major industrial classes, New York City has a higher proportion of workers in white-collar, information-handling occupations than does the United States as a whole. (See Exhibit II).

Advanced communications systems have allowed manufacturing firms based in New York to coordinate production and marketing on a global basis and thus to increase their use of information technology in corporate operations. Even the day-to-day activities of the city's hotels, retail stores, and theaters, rely extensively on new telecommunications systems. Finally, basic telephone service is essential to local shopkeepers (pharmacist, dry-cleaner, and butcher) to respond to customer orders, to make deliveries, and to order new stock. In all sectors of the New York City economy, information transmission systems play a vital role, and the city's future economic health will increasingly depend upon the capacity of private firms to make effective use of advanced telecommunications.

New York as an International Information Capital

New York City's growth as an information capital has been made possible by the advent of telecommunications systems that facilitate both the concentration of financial and headquarters functions in the central business district and the movement of manufacturing and distribution functions. The rich web of face-to-face communications that provide New York firms with access

EXHIBIT II
Proportions of White-Collar and Blue-Collar Workers by Industry, 1984 (United States and New York City)

	White Collar	Blue Collar
Manufacturing		
United States	37.2%	62.8%
New York City	48.0	52.0
Construction		
United States	21.4	78.6
New York City	26.2	73.8
Transportation, Communication and Public Utilities		
United States	49.3	50.7
New York City	56.1	43.9
Wholesale and Retail Trade		
United States	61.8	38.2
New York City	68.9	31.1
Finance, Insurance and Real Estate		
United States	92.9	7.1
New York City	88.5	11.5
Services		
United States	64.3	35.7
New York City	70.7	29.3

Source: U.S. data from U.S. Department of Labor, BLS, *Employment and Earnings,* January 1985. New York City data from Drennan, 1985, derived from Occupational Employment Statistics, New York State Department of Labor.

to the latest information and ideas is closely linked to the new telecommunications technologies of the 1980s. Communications technologies allow firms based in New York to convert new information into profit-making services and to produce decisions that result in the production of goods and services around the world. New York City's headquarters complex includes 62 of the Fortune 500 largest industrial firms, 11 of the nation's 50 largest commercial banks, 10 of the country's top 50 diversified financial companies, and 7 of the nation's largest diversified service companies.[1] In addition, more than 300 foreign banks are located in New York, the largest such center of foreign banks in the United States.

International commerce plays a growing role in the economy of New York City and in the economy of the nation as a whole. In 1972 international trade (the value of imports and exports) represented only 8.9 percent of gross national product (GNP). In 1984 this figure had grown to 18.7 percent of GNP. New York has emerged as the international gateway for overseas communications traffic. In 1981 almost one fourth of all overseas business calls and 15 percent of all overseas residential calls originated in New York City. More than twice as many overseas message units originate in New York City as in Los Angeles, the second leading overseas telephone departure point in the United States. New York City has historically been a leader in international telephone services. The first commercial overseas telephone call was placed from New York City to London on January 7, 1927. Today, the city is the site for new international satellite services between New York City and London.

An Assessment of Telecommunications Systems in New York City

New York City's unmatched set of telecommunications facilities and services has been substantially strengthened through the deregulation of the telecommunications industry. No other city in the world has such a diverse and extensive telecommunications infrastructure, encompassing coaxial cable and fiber optic systems plus major satellite earth stations in outlying parts of the city. The availability of fiber optic systems throughout the city enhances economic development opportunities by information-intensive firms. Further, the planned cable television systems in the boroughs outside Manhattan and proposed digital termination systems (DTS) will constitute significant new elements in the city's telecommunications infrastructure.

New York City's advanced telecommunications infrastructure is a significant asset for economic development because it provides private firms with a wide choice of sophisticated services in a highly competitive environment. Advances in telecommunications technology contribute to the economic productivity of the city's industry by allowing private firms to extend their geographic reach and to market new products and services. For the small- and medium-size firms that characterize the New York City economy, shared tenant services within large office buildings will provide the benefits of state-of-the-art telecommunications technology without requiring large capital investment in equipment and facilities. For the information processing activities that occur in the "back offices" of financial service firms, the extensive telecommunications infrastructure facilitates the decentralization of routine office activities.

The enormous concentration of information-based firms in New York City has led to the development of New York City's unique telecommunications infrastructure. Ken Phillips, vice president of Citicorp, has stated that the Manhattan Central Business District has more than "twice the telecommunications switching capacity of the average foreign country, more computers than a country the size of Brazil, and more word processors than all the countries of Europe combined. Capital investment by business users in private telecommunications systems, communicating word processing systems, computer mainframes, minis, and micros is currently in the billions of dollars and is growing annually."[2]

There are three segments to the telecommunications infrastructure in New York City: (1) long-distance or intercity systems that link New York City with other parts of the nation and world; (2) intracity transmission systems that link telephone central offices, connect subscribers to communications carriers, and/or produce alternative local distribution systems; and (3) local area networks that transmit information within a single building or set of buildings. In light of New York City's role as the nation's leading information center, telecommunications policies and regulations designed to foster competition, including efficient pricing of telecommunications services, are crucial to the continued growth of telecommunications throughout the city.

NEW YORK CITY'S LONG-DISTANCE TELECOMMUNICATIONS SYSTEMS

Long-Distance Fiber Optic Systems

At the regional and national level, fiber optics is gradually replacing satellite and copper wire systems as the transmitter of information at high speeds over long distances. Several characteristics of fiber optic systems enhance their use in telecommunications systems:

1. *Large capacity:* A large amount of information can be transmitted rapidly in a very limited amount of space. The Office of Technology Assessment (OTA) estimates that "a quarter-inch diameter optical cable with two fibers carries as much data as a 3-inch copper cable with 2,000 wires."[3]
2. *Declining cost:* Compared with other telecommunications technologies, the cost per channel of communication over fiber is decreasing rapidly. Within the next three years, the cost of fiber alone will be approximately a few cents per meter. The primary financial constraints are right-of-way costs and installation of fiber and repeaters.[4]
3. *High security:* Fiber is resistant to wiretaps or interference from external sources.
4. *Signal strength:* Fewer repeaters are needed to regenerate signals with fiber than with copper systems, and thus maintenance and installation costs are reduced. Repeaters are needed at 1-mile intervals for copper telephone systems, but only for every 50 miles with fiber systems.
5. *Minimal delay:* Unlike satellite communication, which must travel 23,000 miles to outer space and back, information on fiber travels directly between points. Although the time delay over satellite is not significant for many forms of communication, it represents a major inefficiency

for the integrated digital networks that are increasingly being used for international communications.[5]

Because New York City is the largest single source of national and international telephone traffic, it has become the major hub for current and proposed fiber optic systems in the United States. Presently, there are five long-distance fiber optic communications systems that serve or are planned to serve New York City-based customers. The provision of these long-distance fiber optic services allows users located in the city the benefits of a truly competitive marketplace: choice of vendor, competition in price, and an incentive for innovative services. The major long-distance fiber optic companies serving New York City are described below:

AT&T has, in operation, a fiber optic cable linking New York City, Baltimore, Washington, and Philadelphia.

MCI has a fiber optic cable connecting New York City, Philadelphia, Wilmington, and Washington, D.C. This cable will eventually extend to Miami. In addition, MCI's planned acquisition of Satellite Business Systems (SBS) will allow MCI to incorporate the SBS plan to link New York with Boston.

US Telecom, a subsidiary of United Telecommunications, plans to operate a 23,000-mile fiber network within the next three years. A fiber optic link from New York City to Chicago became operational during the first quarter of 1986.

GTE *Sprint* is completing a fiber optic link from New York City to Sparta, New Jersey, and is constructing a fiber optic link from New York City to White Plains, New York.

Lightnet, a joint venture of Southern New England Telephone and CSX Corporation, is building a 5,000-mile fiber optic network that will serve 43 cities in 24 states east of the Mississippi River. New York City was connected to Philadelphia in 1985, and a route from New York City to Washington and Chicago is to be operational by early 1986. The New York City Lightnet terminal will be located at 60 Hudson Street; two fiber cables will enter New York City, and both are being brought in through an agreement with Teleport Communications, Inc. and the Port Authority of New York and New Jersey. Lightnet is also installing a fiber link for Teleport Communications, Inc. between New York City and Princeton, New Jersey.

Satellite Facilities in New York City

Developed during the 1960s, communication satellites provide a vital link for long-distance communications between New York City and other parts of the nation and world. Communication satellites are accessed by dish-shaped antennas known as earth stations. Large dishes are used for both sending and receiving signals, and smaller dishes are primarily used for receive-only purposes. Commercial satellites have, until recently, used the C-band transmission frequencies between 3.0 and 6.2 GHz. These frequencies, however, coincide with those used for microwave transmission on earth, and this has been a constraint on satellite earth station installations in urban areas already heavily congested with terrestrial microwave traffic, such as New York City. As a result, most satellite facilities serving New York City have been located in Vernon Valley, New Jersey, or in Suffolk County, Long Island, with microwave transmission used to link New York City locations to the satellite earth stations outside the city.

Advances in technology and the desire to avoid this congestion problem have led to increased use of Ku-band transmission frequencies for satellites, which rely on transmission at higher frequencies. The growth of Ku-band satellites has contributed to the installation and operation of new satellite earth stations within New York City by both private users and common carriers. Ku-band satellites, by utilizing a stronger signal, are able to reach smaller satellite dishes, and such satellite dishes can consequently be installed with greater ease on rooftops of office buildings. In addition, the Federal Communication Commission's recent deregulation of a portion of international satellite services has allowed international satellite facilities to be located within New York City. The partial deregulation of international satellite traffic will allow firms located in New York City to obtain the benefits of a competitive marketplace in international telecommunications as well as in domestic services. Given the city's preeminence as a center for international communications, the latest innovations in international services are being developed in New York and other world finance capitals first. The following section will describe the major satellite facilities within the City of New York.

Teleport: A Public-Private Partnership

The largest satellite facilities in New York City are being developed by Teleport Communications, Inc. This represents a public-private joint venture between Merrill Lynch Telecommunications, Inc. and Western Union Communications Systems, Inc., the City of New York, and the Port Authority of New York and New Jersey. The idea for a teleport was based upon the belief that a public entity should provide a facility similar to airports, but for access to communication satellites. Due to the large volume of information flowing into New York City by electronic means the city and Port Authority considered access to communication satellites as vital to the city and region's economy. Microwave congestion reinforced the need for an alternative local distribution system and led to the creation of a fiber optic network linking the teleport on Staten Island to New York City and New Jersey. The 100-acre site on Staten Island was chosen because of its interference-free characteristics plus its access to both New York and New Jersey. In addition to the satellite earth stations to be built, the site will accommodate back office functions plus a telecenter—a 56,000-square-foot building that will control all teleport communications services and provide advanced data and voice services on a 24-hour-a-day basis.

Teleport is designed to serve 17 earth stations; satellite communications started on June 20, 1985, with the inauguration of the Comsat antenna. Portions of Teleport's regional fiber optic network have been in operation since April 1985. Several major firms, including Dow Jones, Bankers' Trust, Citicorp, and Private Satellite Network, are hooked into Teleport's fiber optic cable. In addition, the Catholic Telecommunications Network of America plans to move its broadcast facilities to Teleport and will transmit its programming across the country from Teleport.

Teleport's most distinctive success to date has been as a provider of an alternative means of local communications through its 220-kilometer fiber optic network. AT&T has announced that it will use Teleport's fiber system to link Merrill Lynch to an AT&T office in lower Manhattan, thus bypassing New York Telephone's local network. Teleport's fiber system provides an important intraurban telecommunications system, at relatively low cost, to major users located in the City of New York and surrounding metropolitan region.

Within New York City, Teleport's major service nodes will include 77 Water Street, 2 World Trade Center, 60 Hudson Street, the Empire State Building (350 Fifth Avenue), the Fisk Building (250 West 57 Street), Astoria Studios, 161 Street and Jamaica Avenue, Queens, and the Polytechnic Institute of New York in downtown Brooklyn. In New Jersey, Teleport's major service nodes are in Newark, Jersey City, and the Forrestal Center Campus in Princeton.

Satellite Common Carriers in New York City

There are six companies that operate communication satellite facilities in New York City, providing access to their earth stations on a common carrier basis to major communication users. The common carriers included the following;

Hughes Communication. Now part of General Motors, Hughes Communication operates an earth station from its Spring Creek, Brooklyn, satellite facility. This earth station is linked to Hughes's three satellites, one of which is devoted to video transmission and two of which are devoted to data transmission by private businesses.

ITT Satellite Services. A unit of ITT World Communication Inc., ITT

Satellite offers voice, video, and data transmission to firms located primarily in the lower Manhattan financial district. The Federal Communications Commission has authorized ITT to build an 8-meter, fully digital earth station on the roof of its 60 Broad Street building. This earth station will provide ITT customers with satellite connections between the United States and 200 other nations.

GTE Space Net. Located at Wall Street and Battery Park West, GTE Space Net provides comprehensive services to more than 100 corporations. This earth station communicates with GTE's three SPACENET satellites and two GSTAR satellites.

Private Satellite Network (PSN). PSN is headquartered at 215 Lexington Avenue and develops, installs, and maintains Ku-band satellite systems that provide point-to-multipoint teleconferencing services. PSN is connected to Teleport's fiber optic network and uses the teleport node located at the Empire State Building. PSN provides satellite services to such large firms as J. C. Penney, Merrill Lynch, and Ford Motor Company.

MCI/SBS. Through its planned acquisition of Satellite Business Services, MCI will have access to two SBS earth stations at 1 State Street in lower Manhattan. SBS primarily provides voice transmission by satellite and has extensive ground equipment throughout the United States to provide satellite communications.

International Relay Inc. IRI is the first company to offer a fully digital satellite link from New York City to international locations. IRI has a Ku-band satellite earth station at 345 East 46 Street in Manhattan. Customer access to this site is provided by New York Telephone, Manhattan Cable Television, and by microwave. IRI has a high-speed data link with London, through an agreement with Mercury Communications, Limited. The first user of the IRI-Mercury link was the Associated Press; in addition, the London Stock Exchange provides real-time London stock quotations to New York through the IRI-Mercury link.

Private Earth Stations in New York City

New York City is the capital of the nation's broadcast television industry. These firms make such extensive use of satellite communications systems that it is cost effective to have an earth station devoted exclusively to their operations. For example, NBC maintains and operates an earth station on top of the Celanese Building at 48th Street and the Avenue of the Americas, and ABC-TV maintains and operates an earth station in Manhattan as well.

ISACOMM, a subsidiary of United Telecommunications, has a satellite earth station on the roof of 245 Park Avenue that is used for videoconferencing. Several other firms have their own private earth stations in New York

City. There are also many receive-only satellite dishes used for such purposes as internal corporate data communications, videoconferencing, and access to the Reuters information service.

Public Communication Networks

New York City is also a central link for public communication networks that use packet switching to provide corporations with low-cost domestic communications services. By making maximum use of existing communications facilities, these network services provide a large number of small and medium-size firms access to sophisticated computer-based communications. Packet switching stations operate as a common interface for firms that want to link their facilities. Additionally, these public networks can convert incompatible computer protocols and thereby enhance computer-based communications. The major networking services in New York City include the following:

GTE Telenet. GTE Telenet offers comprehensive services to more than 250 cities and more than 40 countries. In addition, GTE operates a hybrid network service that allows organizations to maintain a private network between commonly used points and access to GTE's public network for intermittently used points. GTE's New York packet switching facilities are located at 1 Penn Plaza.

Uninet, Inc. A subsidiary of United Telecommunications, Uninet provides comprehensive service to 275 domestic metropolitan areas and, through an association with international record carriers, to several overseas locations. Uninet's packet-switching facilities are located at 2 Penn Plaza.

Tymnet. A subsidiary of McDonnell Douglas Company, Tymnet offers comprehensive service to more than 500 domestic metropolitan areas and, through association with international record carriers, to more than 50 countries.

ARGO Communications. ARGO is headquartered in New Rochelle, New York. It uses satellites to provide long-distance domestic network services and uses microwave and leased telephone lines to transmit information within a region. ARGO offers fully digital high-speed data services to 11 other American cities.

NEW YORK CITY'S INTRACITY TELECOMMUNICATIONS SYSTEMS

New York City has more than 5 million telephones. This is the largest concentration of telephones in any American city, and the third largest in the world, after Tokyo and the Paris area.[6] The copper wire linking New

York City's telephones is the only two-way communications link universally available to households and businesses in all five boroughs. The major provider of communications with New York City is New York Telephone, a subsidiary of NYNEX, one of the seven regional holding companies created through the divestiture of AT&T. In addition to copper wire, information is transmitted within New York City by optical fiber, microwave, coaxial cable, and cellular mobile telephone systems.

It is important to recognize that most telecommunications usage is local. Indeed, more than 75 percent of all telephone calls from Manhattan are to points within Manhattan. Similarly, more than 60 percent of the telephone calls from Brooklyn, Queens, Staten Island, and the Bronx are to points within each respective borough. The recent breakup of the city into two area codes, 212 and 718, was designed to reflect in part the localized character of much of the telephone traffic within the city. The large volume of local telephone traffic makes possible and indeed necessary the extraordinarily diverse and dense set of telecommunications facilities and systems.

Fiber Optic Communications within New York City

Fiber optic cable is particularly attractive for use in New York City because it requires relatively little duct space, and underground duct space is a scarce resource. Moreover, fiber is most appropriate to high-volume point-to-point communications, such as connecting telephone switching centers or linking data processing centers. New York Telephone has been quite active in using fiber optics to replace copper wire in its telephone plant. A 48,000 circuit fiber "Ring around Manhattan" (RAM) connects the 12 major switching centers in Manhattan. Ducts have already been laid to increase RAM's capacity by a factor of four if and when demand exists. New York Telephone also has a fiber optic network to provide "Video around Manhattan" that provides Manhattan businesses with enhanced digital services, including wideband video transmission.

In addition, New York Telephone's Interborough Optical Network connects 31 central offices in Brooklyn, Queens, the Bronx, Manhattan, Nassau County, and Westchester County. The Interborough Optical Network consists of 130 miles of cable with a capacity of 5.8 million voice or data conversations and will tie in with New York Telephone's existing fiber network in Manhattan. This system provides the latest telecommunications services to the boroughs other than Manhattan and will enhance not only the telecommunications infrastructure throughout the city, but also the city's economic development potential outside Manhattan.

New York Telephone is one of several fiber optic systems, albeit the largest, within New York City. As discussed above, Teleport Communications, Inc., operates a fiber system connecting Manhattan, Staten Island, Brooklyn, Queens, and New Jersey. In addition, major corporate users have installed their own fiber optic systems within specific areas of Manhattan. Examples of such fiber systems include these:

Manhattan Cable Television and Group W Cable. MCT and Group W use fiber optic trunks in their respective head-ends at 120 East 23 Street and 5120 Broadway.

ITT. ITT operates fiber optic cables in the lower Manhattan financial district. These cables are used as a link to ITT's earth station and to ITT's Long-Distance Center at 1 Whitehall Street.

Citicorp. Citicorp uses fiber optics in its MICRONET system that links its downtown offices (20 Exchange Place, 111 Wall Street, and 55 Water Street) with its headquarters at 399 Park Avenue and Citicorp Center.

McGraw-Hill. There is a fiber optic link between McGraw-Hill's headquarters at 1221 Avenue of the Americas and their satellite offices at 1633 Broadway.

SIAC. The Securities Industry Automation Corporation has combined fiber optic and coaxial cable links between SIAC's information processing center at 55 Water Street and the New York Stock Exchange and between the center and the American Stock Exchange.

Western Union. Western Union provides fiber optic communications for customers and for other communications carriers. The company has the legal authority to use public right-of-way for communications systems, such as fiber optics.

Cable Television Systems in New York City

Two cable television systems are operational in New York City. Both are in Manhattan, but plans are underway for cable television systems to be built in Brooklyn, Queens, Staten Island, and the Bronx. Manhattan Cable Television operates from 86th Street on the east side to 79th Street on the west side, south to Battery Park, and including Roosevelt Island. MCT has approximately 207,000 subscribers and provides 33 channels of cable programming. MCT also has a subsidiary, Manhattan Cable Communications Systems, that provides data transmission service to 12 organizations, including Bankers Trust, the Chase Manhattan Bank, Shearson American Express, Morgan Guaranty, and the City of New York.

Manhattan Cable Television data transmission services are not distance sensitive and thus provide an advantage to users between lower Manhattan and midtown Manhattan. Once regarded as unreliable for data transmission, cable television systems have been upgrading their plant, and Manhattan Cable Television has installed a redundant transmission system in the central business district. Furthermore, MCT will soon be linked to the New York and American Stock Exchanges.

Group W Cable operates in upper Manhattan, has approximately 105,000 subscribers, and provides 27 channels of cable programming. Because Group W Cable is located outside the Manhattan Central Business District, its services are primarily limited to providing residential customers with cable television.

An important example of how coaxial cable can be used for internal corporate communications is provided by the Chase Manhattan Bank's "Metronet" system in lower Manhattan. Chase Manhattan uses coaxial cable to link four buildings—59 Maiden Lane, 80 Pine Street, 1 New York Plaza, and Chase Manhattan Plaza—for high-speed data transmission among these facilities. In addition, Chase Manhattan uses the coaxial cable for video conferencing between New York Plaza and Chase Manhattan Plaza and plans eventually to use the system for voice communication. The Chase "Metronet" system is connected via Manhattan Cable Television's data transmission system to Chase Manhattan facilities at 350 Park Avenue as well.

Terrestrial Microwave Systems in New York City

Microwave transmission systems are used for point-to-point communications. Transmission occurs over a straight line, and this requires a line-of-sight transmission path. Users of microwave technology include broadcast television stations for their studio-to-transmitter links, AT&T, New York Telephone, RCA, and large commercial banks. The following are among the users of microwave transmission.

Hughes Communication links its Manhattan-based customers to its satellite earth station in Brooklyn.

Broadcast Television Stations use microwave to connect studios to transmitters atop the World Trade Center. Mobile television vans use microwave to transmit "live news" to television studios, and WNET also uses microwave to link its Newark, New Jersey, and Manhattan studios.

Citicorp uses microwave transmission to communicate between facilities within New York City and to connect with its satellite facilities in New Jersey.

Manufacturers Hanover uses microwave transmission to link its corporate headquarters at 270 Park Avenue with its computer center at 4 New York Plaza and back office operations located at Hicksville, Long Island.

McGraw-Hill has a microwave link from its headquarters in Manhattan to its corporate data processing and warehouse facilities in Heightstown, New Jersey.

The New York Times uses microwave to transmit "news copy" to its printing presses in New Jersey and to satellite earth stations for distribution to printing presses across the country.

RCA operates a private line voice and data service from its central terminal office at 50 Broad Street to its earth station in Vernon Valley, New Jersey.

American Satellite Corporation has a microwave link from the World Trade Center to its earth station in Vernon Valley, New Jersey.

Eastern Microwave is a common carrier microwave company that provides microwave transmission to a variety of private and public users. It has microwave relays at the World Trade Center, Penn Plaza, the Gulf and Western Building on Columbus Circle, and at other locations in New York City.

At the present time, it is virtually impossible to obtain a low-frequency (and more reliable) microwave transmission path in Manhattan. Either a user must do a search of FCC records and of actual frequency routes before an application is made to the FCC, or a user may lease a frequency from a regional common carrier (such as Eastern Microwave) that has been assigned groups of frequencies.

Cellular Mobile Radio in New York City

Until recently, mobile telephone service in the United States was highly limited. The advent of cellular mobile radio has created a new environment for mobile telephone service, and the Federal Communications Commission has authorized two firms to provide cellular mobile service in New York City, NYNEX and Cellular Telephone Company. NYNEX currently has more than 10,000 subscribers and expects to have 110,000 customers by 1990. The Cellular Telephone Company's system was given FCC authorization in October 1984.

Smart Buildings and Local Area Networks in New York City

A smart building can refer to: (1) an integrated management system for elevators, energy, security, and other building services; (2) an integrated telecommunications network for local, long-distance, and enhanced services (such as voice mail and teleconferencing); or (3) a building that provides integrated telecommunications and building management services. A local area network (LAN) provides a communications link, within a single building or among a number of buildings, for personal and mainframe computers, data storage banks, printers, and other computer and telecommunications equipment. The LAN can be provided by coaxial cable, copper telephone wire, or fiber optic cable, depending on the particular design and use.

Smart buildings usually offer "shared tenant services" (STS), sophisticated telecommunications services that are marketed to building tenants and that provide economies of scale and one-stop convenience for small- and middle-size firms. For real estate developers, shared tenant services offer a service to tenants and a potential source of revenue; in most cases, developers have formed partnerships with telecommunications firms to market and manage such building-based communications services.

Tishman-Speyer is constructing a smart building at 375 Hudson Street in lower Manhattan that will be designed to accommodate the latest telecommunications and computer systems. There will be facilities for satellite transmission from the roof, large floor areas for computer facilities, and vertical duct space for advanced transmission systems. Several other office buildings are under construction in New York City with similar capabilities for shared tenant services.[7]

TELECOMMUNICATIONS AND ECONOMIC DEVELOPMENT

From an economic development perspective, the City of New York is vitally interested in the telecommunications market and infrastructure. Information intensive private firms require access to sophisticated telecommunications services and facilities that are competitively priced. To a large extent, continued information sector growth will be dependent on meeting user demands. To date, the high demand for specialized telecommunications services has enabled New York to take the lead in introducing competitive telecommunication services.

Continued growth and innovation will depend upon regulatory policies that contribute to New York City's role as a world center for information-based services. Conversely, economically inefficient regulatory policies could constrain market competition and technological innovation and limit the continued development of the city's telecommunications infrastructure. One major policy issue concerns the growth of competitive technologies that allow large-volume telecommunications users to bypass the public network operated by New York Telephone. Organizations have always been able to bypass the public network (often through services provided by New York Telephone), but the availability of new telecommunications systems and the desire to avoid long-distance access charges have contributed to increased use of bypass technologies. The Federal Communications Commission defines *bypass* as "the transmission of long-distance messages that do not use the facilities of local telephone companies available to the general public, but that could use such facilities."[8]

It is useful to distinguish between "economic" and "uneconomic" bypass. "Uneconomic" bypass refers to the use of alternative communications facilities whose actual *costs* are below the telephone company's *rates,* but that may still be in excess of the telephone company's underlying costs. By contrast, "economic" bypass involves genuinely more efficient telecommunications service by alternative means.

A widely held concern is that new telecommunications systems will permit "uneconomic bypass" by encouraging providers to enter the market primarily to serve businesses with high telecommunications costs located in the central business district and draw them away from the public telephone network. This has serious implications for the financial base of the telephone network, since ⅓ of New York Telephone's long-distance revenues are generated by three tenths of 1 percent of its business customers.[9] New York Telephone reports that one in three of its largest customers already engages in bypass and that more than half of New York Telephone's largest customers plan to bypass in the future.[10] For large users that do bypass, however, many still rely on New York Telephone for a substantial part of telecommunication service.[11]

■ CONCLUSION ■

The intense concentration of information handling and processing activities in the City of New York has given rise to a sophisticated and diverse telecommunications infrastructure that is unmatched by any city in the United States. Technological advances in conjunction with the deregulation of the telecommunications industry have strengthened the city's information infrastructure and possibilities for economic growth. New communications technologies enhance the productivity of the city's industries by allowing firms to extend their geographic reach and to market new products and services on a global basis. This is essential to the growing international trade that is centered in the world's "gateway" cities.

Although the teleport project on Staten Island has received a great deal of public attention, this chapter has demonstrated that the telecommunications infrastructure in the City of New York consists of a diversity of telecommunications systems that both complement and compete with the services provided by Teleport Communications, Inc. Moreover, it highlights the important role of the private sector in planning and developing new telecommunications systems that encompass fiber optics, microwave, coaxial cable, and satellite, as well as twisted pairs of copper wire. Finally, this chapter provides an important case study of the relationship of teleports to the changing telecommunications infrastructure within cities and the need to understand the system of wires, ducts, and channels that transmit information within large cities.

■ NOTES ■

1. *Fortune,* April 29, 1985, and June 10, 1985.

2. Kenneth L. Phillips, "Telecommunications and New York in the Year 2000," Testimony before the New York City Commission on the Year 2000, May 9, 1985, p. 3–4.

3. Congress of the United States, Office of Technology Assessment, *Information Technology R & D: Critical Trends and Issues* (Washington, D.C.: U.S. Government Printing Office, February 1985), p. 67.

4. A. M. Rutkowski, "Satellite Competition with Optical Fiber," paper presented at Satellite Summit, April 1985.

5. Ibid, p. 3.

6. *The World's Telephones: A Statistical Compilation as of January 1983.* (Morris Plains, N.J.: AT&T Communications), 1985.

7. Eric N. Berg, "Sharing Telephone Services," *The New York Times,* August 27, 1985, p. D1.

8. Federal Communications Commission, Common Carrier Bureau, *Bypass of the Public Switched Network,* December 19, 1984, p. 7.

9. Michael Wines, "Teleports May be the Newest Threat to Bell Companies' Local Dominance," *National Journal,* November 12, 1983, p. 2352.

10. Supplemental Testimony of Dr. Joseph S. Kraemer, New York Public Service Commission, Case 28710, June 1, 1984, p. 2.

11. Mitchell L. Moss, "Telecommunications Policy Issues for New York State," Report to the Governor's Council on Fiscal and Economic Priorities, December 1985.

Author

Mitchell L. Moss
New York University
New York, New York

Mitchell Moss is associate professor of planning and public administration at New York University and the deputy to the chairman of the New York State Council on Fiscal and Economic Priorities. He served as chairman of NYU's Interactive Telecommunications Program from 1981 through 1983.

Professor Moss currently directs a project on the impact of new communications technology on cities. He was principal investigator for the National Science Foundation-sponsored experiment with two-way cable television conducted in Reading, Pennsylvania, and has also directed projects for the Corporation for Public Broadcasting, Benton Foundation, and U.S. Department of Commerce. His articles have appeared in *Telecommunications Policy, Urban Affairs Quarterly, Journal of Urban Law,* and *Journal of Communications,* and he is the editor of *Telecommunications and Productivity* (Addision-Wesley, 1981).

Moss has served as a consultant on telecommunications and economic development to the City of New York, Port Authority of New York and New Jersey, and New York State Energy Research and Development Authority. He is a member of the New York State Urban Development Corporation's High Technology Advisory Committee. Moss received a B.A. from Northwestern University, M.A. from the University of Washington, and Ph.D. from the University of Southern California.

This article is based on a report prepared for the Office for Economic Development, City of New York. Any findings expressed in this article are those of the author and do not necessarily represent the views of the sponsoring agency. The author wishes to acknowledge the valuable research assistance of Mr. Andrew Dunau.

Index